Lecture Notes of the Institute for Computer Sciences, Social-Informatics and Telecommunications Engineering   12

Cristian Hesselman   Carlo Giannelli (Eds.)

Mobile Wireless Middleware, Operating Systems and Applications – Workshops

Mobilware 2009 Workshops
Berlin, Germany, April 2009
Revised Selected Papers

 Springer

Volume Editors

Cristian Hesselman
Novay, P.O. Box 589
7500 AN Enschede, The Netherlands
E-mail: cristian.hesselman@novay.nl

Carlo Giannelli
University of Bologna, c/o Ingegneria - DEIS
Viale Risorgimento, 2
40136 Bologna, Italy
E-mail: carlo.gianelli@unibo.it

Library of Congress Control Number: 2009932177

CR Subject Classification (1998): C.2.1, C.2.4, D.1.3, D.2.12, D.4.2, D.4.7, E.1, H.3.4

ISSN   1867-8211
ISBN-10   3-642-03568-X Springer Berlin Heidelberg New York
ISBN-13   978-3-642-03568-5 Springer Berlin Heidelberg New York

springer.com

© ICST Institute for Computer Science, Social-Informatics and Telecommunications Engineering 2009
Printed in Germany

Typesetting: Camera-ready by author, data conversion by Scientific Publishing Services, Chennai, India
Printed on acid-free paper   SPIN: 12730551   06/3180   5 4 3 2 1 0

Preface

Software systems for wireless and mobile communications are a key component in pervasive computing and are crucial for the materialization of easy-to-use and intelligent services that people can use ubiquitously. As indicated by its acronym (*MOBILe* Wireless Middle*WARE*, Operating Systems, and Applications), these are the type of systems that form the topic of the MOBILWARE conferencing series. In particular, the goal of MOBILWARE is to provide a forum for researchers and practitioners to disseminate and discuss recent advances in software systems for wireless and mobile communications, ranging from work on communication middleware and operating systems to networking protocols and applications.

For its second edition, held in Berlin in April 2009, the MOBILWARE Organizing Committee decided to add a full day of workshops on topics related to the main conference. Our goals were threefold:

1. Put together a high-quality workshop program consisting of a few focused workshops that would provide ample time for discussion, thus enabling presenters to quickly advance their work and workshop attendees to quickly get an idea of ongoing work in selected research areas.
2. Provide a more complete picture of ongoing work by not only including technical workshops, but also workshops on business and user aspects. We expected that this multi-viewpoint approach would be an added value as technology, business models, and user experiences are usually interrelated.
3. Create a breeding ground for submissions for MOBILWARE 2010 and beyond.

Our call for workshop proposals reflected these goals and attracted a total of six high-quality proposals. All proposals were written by international teams of different organizations, including teams consisting of researchers from academia and industry, which suggested a wide interest in our goals and approach.

After a thorough review process, we accepted four workshops, two of which were purely technical, one of a mixed technical/end-user nature, and one focusing on business models:

- *Ubi-Islands 2009*, the First International Workshop on Interconnecting Ubiquitous Islands Using Mobile and Next Generation Networks
- *WASP 2009*, the First International Workshop on Wireless Sensor Networks Architectures, Simulation, and Programming
- *UCPA 2009*, the First International Workshop on User-Centric Pervasive Adaptation *BMMP 2009*, the First International Workshop on Business Models for New Mobile Platforms

As the MOBILWARE Workshop Chair, I am happy and proud to say that we can look back on a very successful and smooth workshop day, which took place on April 27, 2009. For example, the Ubi-Islands workshop was well attended by about 25 participants and received very positive feedback throughout. Everyone enjoyed a lively and well-focused forum with highly inspiring talks and discussions, as was the case at UCPA 2009. BMMP 2009 was proud to gather the academic and business research community dealing with business models for mobile platforms for the first time. They devoted specific sessions to platform theories and concepts, mobile platform business models, and two-sided market models for mobile platforms. BMMP involved high-quality debates and submitted selected papers to a special issue of *Telematics & Informatics*.

On behalf of the MOBILWARE Organizing Committee I would like to thank Andreas Fasbender (Ubi-Islands 2009), Christian Kroiß (UCPA 2009), Pieter Ballon (BMMP 2009), and Soledad Escolar Díaz (WASP 2009) for their excellent coordination work and making their workshop a success. I would also like to thank Paolo Bellavista (MOBILWARE General Co-chair), Gergely Nagy (Conference Organization Chair), and Carlo Giannelli (MOBILWARE Publication and Web Chair) for their continuous support sin organizing the MOBILWARE 2009 workshops.

Next year's edition of MOBILWARE will be held in Chicago in the USA, from June 30 until July 2, and it will again be preceded by a day of workshops. I am very much looking forward to working with everyone again to also make the MOBILWARE 2010 workshop program a success.

Cristian Hesselman

Organization

Steering Committee

Imrich Chlamtac (Chair)	Create-Net, Italy
Paolo Bellavista	University of Bologna, Italy
Carl K. Chang	Iowa State University, USA

General Co-chairs

Paolo Bellavista	University of Bologna, Italy
Linda Xie	University of North Carolina at Charlotte, USA

Technical Program Committee Co-chairs

Jean-Marie Bonnin	ENST, France
Thomas Magedanz	Fraunhofer FOKUS, Germany

Conference Organization Chair

Gergely Nagy	ICST

Conference Publicity Co-chairs

Europe: Andrej Krenker	Sintesio, Slovenia
North America: Jatinder pal Singh	Stanford University, USA
North America: Janise McNair	University of Florida, USA
South America: C. Esteve Rothenberg	CpQD, Brazil
Middle East: Ghadaa Alaa	Information Technology Institute, Egypt
Asia: Michael Chen	III INSTITUTION, Taiwan

Publication and Web Chair

Carlo Giannelli	University of Bologna, Italy

Workshop Chair

Cristian Hesselman	Telematica Institute, The Netherlands

Local Chair

Julia Ovtchinnikova   Fraunhofer FOKUS, Germany

BMMP Workshop Organizers

Pieter Ballon   IBBT & Vrije Universiteit Brussel, Belgium
Harry Bouwman   Delft University of Technology, The Netherlands
Timber Haaker   Telematica Institute, The Netherlands

WASP Workshop Chairs

Jesús Carretero   University Carlos III of Madrid, Spain
Soledad Escolar   University Carlos III of Madrid, Spain

WASP Workshop Program Committee

Hussein Abdel-Wahab   Old Dominion University, USA
Paolo Bellavista   Bologna University, Italy
Azzedine Boukerche   University of Ottawa, Canada
Saad Biaz   Auburn University, USA
Alejandro Calderón Mateos   University Carlos III of Madrid, Spain
Claudia Canali   University of Modena, Italy
Jose M. Dana   CERN, Switzerland
Mostafa Hashem Sherif   ATT Labs. USA
María Cristina Marinescu   University Carlos III of Madrid, Spain
Teresa Olivares   University of Castilla-La Mancha, Spain
Luis Orozco   University of Castilla-La Mancha, Spain
Antonio Corradi   University of Bologna, Italy
Douglas Reeves   North Carolina State University, USA
Vlad Olaru   Cluj-Napuca University, Romania

UCPA Workshop Organizers

Nikola Serbedzija   Fraunhofer FIRST, Germany
Martin Wirsing   LMU Munich, Germany

UCPA Workshop Program Committee

Alois Ferscha   Johannes Kepler Universitat Linz, Austria
Andreas Schröder   LMU Munich, Germany
Nikola Serbedzija   Fraunhofer FIRST, Germany
Mladen Stanojevic   IMP, Serbia
Martin Wirsing   LMU Munich, Germany
Franco Zambonelli   Universita di Modena e Reggio Emilia, Italy

Ubi-Island Program Committee

Frank Reichert	University of Agder, Norway (Chair)
Andreas Fasbender	Ericsson GmbH, Germany
Frank den Hartog	TNO, The Netherlands
Hiroyuki Morikawa	University of Tokyo, Japan
R. Venkatesha Prasad	TU Delft, The Netherlands
Johan Hjelm	Nippon Ericsson KK, Japan

Table of Contents

BMMP Workshop

WASP Workshop

UCPA Workshop

Ubi-Islands Workshop

"Just Another Distribution Channel?"

Wolter Lemstra[1], Gerd-Jan de Leeuw[1], Els van de Kar[1], and Paul Brand[2]

[1] TUDelft, Department Technology, Policy & Management,
Jaffalaan 5, NL-2628 BX Delft, The Netherlands
[2] Stratix Consulting, Villa Hestia, Utrechtseweg 29
NL-1213 TK Hilversum, The Netherlands
w.lemstra@planet.nl

Abstract. The telecommunications-centric business model of mobile operators is under attack due to technological convergence in the communication and content industries. This has resulted in a plethora of academic contributions on the design of new business models and service platform architectures. However, a discussion of the challenges that operators are facing in adopting these models is lacking. We assess these challenges by considering the mobile network as part of the value system of the content industry. We will argue that from the perspective of a content provider the mobile network is 'just another' distribution channel. Strategic options available for the mobile communication operators are to deliver an excellent distribution channel for content delivery or to move upwards in the value chain by becoming a content aggregator. To become a mobile content aggregator operators will have to develop or acquire complementary resources and capabilities. Whether this strategic option is sustainable remains open.

Keywords: Mobile operators, content providers, industry structure, business models, technology convergence, strategic options.

1 Introduction

In the late 1980s and early 1990s mobile telecommunications became synonymous with "a license to print money". The introduction of competition in the mid 1990s challenged this perspective, but due to the spectacular growth in demand the market could easily accommodate multiple providers. Moreover, innovation such as the use of prepaid cards unlocked new market segments, the younger generation and the consumers at the so-called 'bottom of the pyramid' [1].

More recently the profitability of the mobile business model has come under pressure due to, e.g., intensity of competition, market saturation and license fees. One response has been to expand the service portfolio from voice only to voice and data. The unplanned but very welcome success of SMS provides a leading example. However, SMS resembles voice as the messages are individually charged. The data-driven business model as we have come to appreciate it in the fixed network domain,

C. Hesselman and C. Giannelli (Eds.): Mobilware 2009 Workshops, LNICST 12, pp. 1–12, 2009.

is characterised by a flat fee depending on the data rate. While the prospect of growing data volumes is attractive to the mobile operator the prospect of becoming a 'bit pipe' provider has less appeal. Ideally a sustainable competitive business model is built on some aspect of unique resources or capabilities, or some aspect of 'captive audience' [2].

An alternative strategic avenue that is being pursued by mobile operators is to move up in the value chain by offering value-added services. The technological developments from analogue to digital encoding of information and the use of the TCP/IP protocol stack as part of the Internet has made this transition possible. This has resulted in a plethora of academic contributions on the design of new business models and service platform architectures, e.g., Van de Kar [3], Bouwman, De Vos, et al. [4], Bouwman, Zhengjia, et al. [5] and Ballon [6]. However, a discussion of the challenges that operators are facing in adopting these business models is lacking.

The purpose of our research is to explore these challenges through assessing the impact of convergence on the business models in the telecommunication and content industry. The research is aimed at assessing the strategic options for the mobile operator as part of the value network of the content provider. Thereby assessing the attractiveness for a mobile operator of becoming a content aggregator. In our research we apply a comparative, longitudinal approach.

This contribution is organised as follows: first, we provide a short description of the business model development in the mobile telecommunication, film and television industry. We identify the core resources and key capabilities of each business model and assess how the various assets can become complementary in providing mobile content services. In section 3 we assess the integrated value network that evolves as a result of converging technologies. In section 4 we discuss the strategic options for the mobile operator within the content industry. In section 5 we provide our conclusions.

2   The Development of the Industries

Although technological convergence creates a cross-over between the once separate telecommunications and content industries, the business models are distinct, based on the development of totally different sets of resources and capabilities.

Telecommunication Industry Development – Business Model Evolution
The (mobile) telecommunications provider has a unique set of resources and capabilities. On the one hand there are the infrastructure assets and on the other hand the direct customer relationships, which includes a billing relationship. In the business model of the telecommunications industry the operator has no involvement in the content, in fact by law operators are obliged to transfer the communication transparently, without tampering with the content. Fig. 1 reflects the related value chain in its basic form. The communication (C) is two-way symmetrical, as it takes place in two directions, and is one-to-one, i.e., between the A- and B-subscriber. Typically both parties have a subscription for the use of the network.

The telecommunications business model has been optimized since the invention of the telephone in 1876, resulting in the so-called voice-driven circuit-mode paradigm.

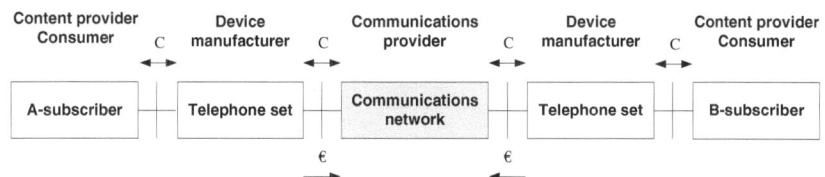

Fig. 1. Value chain of the telecommunications provider

Key attributes of this paradigm are vertical integration, the charging based on time, duration and distance; and the intelligence being located within the network.

With the ARPA[1] developments starting in the 1950s a new data-driven packet-mode paradigm has emerged to become epitomized by the Internet and characterized by the use of the TCP/IP protocol that allows the decoupling of applications from the underlying communications network. In principle all types of applications voice, data, video, images can be supported by all types of networks using twisted pair copper cable, coax, optical fibre, or radio waves. Key attributes of this paradigm are a horizontal structure, a flat-fee charge based on the data rate provided, and the intelligence being located in the terminals connected to the network.

The packet-mode paradigm is now subsuming the circuit-mode paradigm as telecommunications providers are migrating the core network infrastructure to become All-IP. The mobile operators started the transition with GPRS in 1999 now rapidly transitioning towards HSPA, providing packet access in the Mbit/s range.

As part of the business model the mobile operator owns a unique combination of resources and capabilities to serve the 'communications consumer on the move', i.e., the mobile network and operational infrastructure. Moreover, the consumer typically has one mobile subscription and, hence, represents a captive audience while on the move. The customer relationship includes billing. Operators that own and exploit both fixed and mobile network infrastructures have the benefit of a more extended captive audience, both on the move and at home.

Film Industry Development – Business Model Evolution

The content industry emerged with the film and the cinema in the 1890s. The generic value chain is depicted in Fig. 2 [7]. This model is one way and one-to-many: the film distributor initiates the production of the film and uses a distribution channel (enabled by physical transport) to provide the film to the cinemas, where the consumers come together to view the content [7, 8].

The proceeds from the viewers flow back, via the cinema operators to the distributors. In this business model the film distributor carries the production risks. The film distributor distributor has a strong relational network in the world of film production and deep knowledge about the demands and tastes of the consumer. The film distributor may own the film studio, otherwise the distributor has become very light on physical assets as the cinemas are mostly independently owned.

---

[1] ARPA: Advanced Research Projects Agency; US Government agency through which the early developments that have led to the current day Internet were funded.

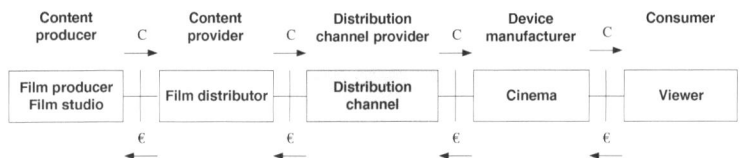

Fig. 2. Value chain of the content provider

Television Broadcast Development – Business Model Evolution

In the 1930s a new content industry segment emerged with the introduction of television, as a next step in the evolution from radio broadcasting in the 1920s [9]. Through the use of powerful radio transmitters content is broadcast to consumers. For the first time consumers can stay at home to watch content, films as well as TV-programs initiated and funded by the television broadcasters. The generic value chain model of the TV-broadcaster is depicted in Figure 3.

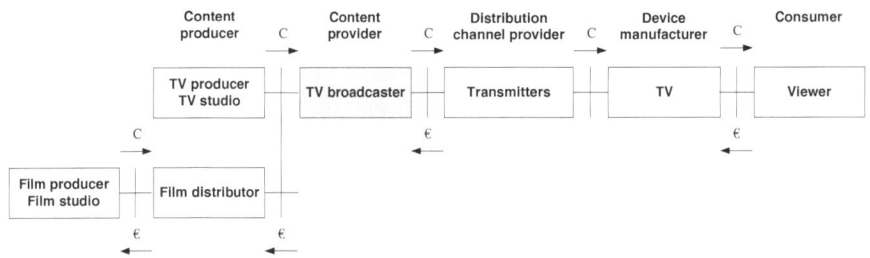

Fig. 3. Value chain of the TV broadcaster

In the Netherlands, as an example of the development in Europe, the government provides licenses to broadcasters with the objective to provide a 'pluriform' programming. Consumers initially paid for a RTV-license, the proceeds were used to fund the broadcasters and the program production. In the late 1980s commercial broadcasters operating from abroad forced a change in the regulatory regime, resulting in TV advertising being allowed and commercial broadcasters being permitted to operate within the Dutch jurisdiction [10].

The TV becomes an alternative distribution channel for the film industry and film is a source of premium content for the TV broadcaster. The royalty payments are determined by the reach of the broadcaster and the viewing rate.

In the 1970s the video cassette recorder (VCR) is being introduced. Next to new functionalities, this device introduced a new distribution channel for the film producer and represents a 'bypass' channel for the TV-broadcaster and the cinema operator.

For TV-users outside the range of a domestic transmitter or TV-users interested to receive foreign programs the launch of satellite services provides for an alternative distribution channel since the 1970s. In essence satellite services are provided according to the business model of a TV-broadcaster, using a different distribution channel requiring a different antenna and the use of a set-top box [9].

Through the set-top box Pay-TV becomes a new element in the business model. The satellite broadcaster has the opportunity to aggregate both film content and TV-content from different sources and may initiate the production of its own content, e.g. procure the rights to 'air' sports events, to be bundled in a branded channel (viz. Canal Plus and Sky). The satellite broadcaster has the benefit of having a captive audience with a large geographical coverage, which through the set-top box can be segmented geographically.

An important change in the terrestrial distribution channel occurs with the emergence of cable systems for the local distribution of the RTV-signals. In the 1970s these municipal CA-TV systems started to emerge in the Netherlands [10]. CA-TV operators introduced a new element in the business model of TV as they can aggregate channels and select which channels will be passed on for further distribution. By Dutch law the cable operators have an obligation to pass through the public broadcasting channels and are required to provide a minimum number of channels.

Depending on the reach and the type of program the operator may have to pay the film producer or TV-broadcaster for the right to pass on a movie or program.

With the introduction of digital transmission on the cable systems combined with the use of a set-top box the cable broadcaster can offer multiple bundles of channels, which may be local or regional and provided for free or for a fee. In this digital case the consumers are representing a captive audience.

Each of the broadcasters has a unique set of resources and capabilities. Albeit, the TV-broadcaster may only have a license to use transmitters at certain locations and at certain periods of time. The cable broadcasters typically own and exploit the cable infrastructure. The satellite broadcasters lease transponder capacity at satellites owned by consortia of telecom or broadcasting operators.

In terms of relational assets the broadcasters have strong networking relationships with the TV-producers, film distributors and other providers of (premium) content. Through viewing statistics and viewing panels the broadcasters obtain valuable insights in the preferences of the consumers. Programs are bundled and channels offered to meet the needs of specific consumer segments. By providing a portfolio of programmes for a fee, the cable and satellite broadcasters obtain direct feedback on consumer preferences.

3 Analysis of Technological and Industry Developments

The Internet becoming accessible to general public in the beginning of the 1990s [11] forms the start of the technological convergence of the telecommunications and the content providing industries. This convergence is taking place at the 'distribution channel' and at the 'device' level. The result of these developments is depicted in Fig. 4 in the form of a value network for content provision. In the following sections we discuss the impact of technological convergence on three aspects in the value network: the distribution channel, content provision, and devices.

Distribution Channel. With the transition from analogue to digital encoding of all types of information (voice-data-image-video) and the introduction of packet-mode communication using the TCP/IP protocol stack technologically different infrastructures

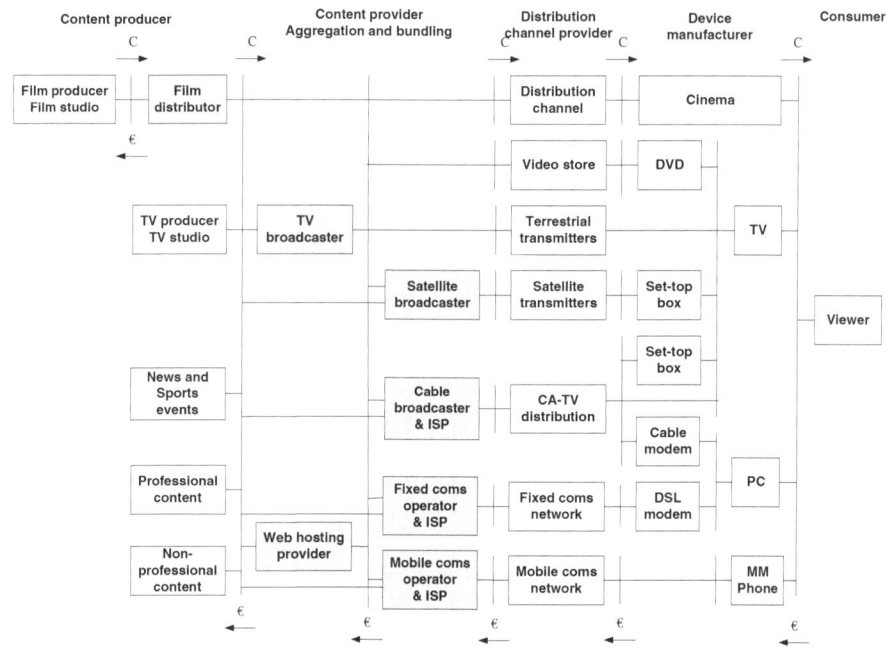

Fig. 4. Value network for content provision

are able to provide the same type of services. The telecommunications networks supporting packet-mode communication, both fixed and mobile, can stream digital RTV-signals using IP. By upgrading cable networks to two-way communication they can provide next to RTV signals, broadband Internet access, and voice communication through IP (VoIP).

Both types of communication networks can in principle support the same categories of services. Albeit, the difference in underlying infrastructures leads to difference in the attributes of the services being provided. For instance the telecommunications networks are not optimized for broadcasting and, hence, are capacity limited in conveying TV-signals in real-time. The use of peer-to-peer protocols, such as Bittorrent, is aimed at alleviating this problem.

This difference also explains the strong interest of, e.g., KPN the Dutch incumbent telecommunications provider to obtain a DVB-T license to provide digital terrestrial RTV-broadcasts.[2] While the service is offered on a stand-alone basis, it provides KPN an opportunity to offer 'triple-play' services in competition with the cable broadcasters. KPN also exploits the DVB-H network for mobile TV.[3] As KPN also has a mobile communications branch, it can provide in principle 'quadruple-play' services, an offer which the cable broadcasters can as yet not match.

---

[2] DVB-T: Digital Video Broadcasting-Terrestrial. Standard for digital broadcasting of RTV signals using radio transmitters.

[3] DVB-H: Digital Video Broadcasting-Handheld. Based on DVB-T, optimized for handhelds.

As a result of these technological developments cable broadcasters and communications providers have become competitors in converging markets.

Content Provision, Aggregation and Bundling. Through the Internet, professional as well as non-professional entities have the opportunity to distribute content to consumers. This distribution is facilitated through application hosting organisations. YouTube is an example of content hosting for non-professionals.

The wealth of information also confuses the user who may be enticed to receive content in a more convenient manner. The BBC iPlayer is an application that provides easy access to content from the BBC. The initiative by Apple Inc. to create the iTunes web store is aimed at creating an eco-system around its own platform of products and represents a closed system for content retrieval (music, games, video) at a (reasonable) fee. Thereby Apple assumes the role of professional content hosting provider.

These different models reflect the battle between open and closed business models for content provision, whereby the latter may be more attractive to content providers who invest in content production and distribution for-a-profit.

Devices. The use of the Internet is enabled though intelligent terminals, such as the PC and laptop, as well as mobile devices such as PDAs and multifunctional mobile phones. All these devices can be used to view video content. Which device is being used at a certain point in time is largely a matter of convenience, linked to the desired degree of mobility by the consumer and the perception of quality. The form factor of the device plays also a major role, as is the mode of viewing (primary, back-ground).

Interestingly, computers have also a large storage capability, which provides for a time-shift function, hence, content can be downloaded and stored for later viewing. This implies a replacement for the VCR and to a lesser extent a replacement for the more recently introduced DVD.

Results of Technological Convergence. The result of technological convergence is a multiplicity of distribution channels being available for content providers to reach the consumer. Moreover, consumers have access to content through multiple devices.

The value of the content being provided determines which (combination of) distribution channel(s) is used by the content providers. The model that has emerged suggests that new films are first distributed to the cinemas and subsequently provided for TV viewing and ultimately made available on DVD. Also a broadcasting model is emerging whereby news and sports are first provided in real-time through the traditional broadcasting networks, to be made available subsequently for delayed viewing via the Internet.

The multiple distribution channels suggests that consumers will be able to find and access any type of content that is either in the public domain or accessible for a fee in the private domain of content providers. The traditional role of selection and bundling of programs by broadcasters would seem to become redundant and challenged by new content re-distribution initiatives, such as Joost (which is web based, using a Popcorn Hour set-top box in combination with advertising).

4 Discussion: Position of a Mobile Operator in the Content Industry

In this section we come back to the main research question: what are the options for mobile communications provider facing an industry trend towards content?

The starting point for the discussion is that the technological changes that have affected the fixed communication network (from circuit-mode to packet-mode) will affect the mobile communication sector in a similar manner; resulting in a change in the revenue model from unit fee to flat fee and a change in the service delivery model from vertically integrated to a more open and diverse model. This calls for a strategic re-orientation by the mobile communications provider.

Management literature suggests that strategy should be aimed at achieving a good fit between the firm and the environment. This may be achieved by the firm adapting to the environment (applying an *outside-in* approach) or the firm exploiting its resource base to shape the environment (applying an *inside-out* approach). Adaptation to the market may require reshaping or extension of the resource base [2]. Moreover, Porter argues that a firm has three generic options to achieve strategic advantage: (1) strive for overall cost leadership across the industry; (2) applying differentiation in delivering uniqueness perceived by the customer; (3) focussing on a particular market segment [12]. Reviewing the position and perspective of the five major actors in the industry will lead us to conclude on the options available to the mobile operator.

Mobile communication provider perspective
The strategic position of the mobile operator has improved with the market penetration of mobile becoming higher than fixed. Moreover, mobile is not only an extra communication channel, but increasingly a substitute for fixed lines.

Constraining the mobile operator position in content provision is the design of the communication networks as two-way and one-to-one. Broadcasting networks, on the other hand, are designed for one-way, one-to-many content distribution. While special protocols, such as Bittorent, make content distribution in the Internet more feasible, as yet they do not allow for large scale cost effective real-time distribution.

An alternative terrestrial communication infrastructure is available for digital broadcasting using DVB-H, which makes TV-broadcasting to handhelds possible, albeit with a limited number of channels. Moreover, the availability of DVB-H is constrained due to limited availability of licenses, which is due to limited availability of RF spectrum. See also Braet and Ballon [13], and Curwen and Whalley [14].

The need for convenience suggests that consumers will require DVB-H to be combined with mobile functionality in one handset. This again suggests a greater interdependency between the network providers and the device manufacturers.

For mobile providers to become a `distribution channel of choice` for providers of mass market content they will have to find more effective ways to distribute content to consumers. Three principle options are open: (1) to stimulate the industry at large to develop effective multicast protocols within the mobile networks; (2) upgrade the network to higher data volume capacity, allowing large numbers of simultaneous users to receive the same content on a 'unicast' basis; and (3) to use a separate broadcasting technology such as DVB-H.

The implementation of option (1) resides within the realm of the mobile communications industry and the parties involved are already aligned towards that (longer term) goal; however, the potential for multicast in cellular networks is limited. Option (2) is part of the development path towards 4G (LTE), but may not be sufficiently scalable [15]. For the implementation of option (3) the mobile provider is dependent on the regulator. Considering that the regulator (or policy makers) may perceive RF spectrum as a scarce resource, they may wish to distribute RF capacity equitably and/or reserve spectrum for new entrants. This may result in the exclusion of established mobile providers from obtaining a DVB-H license. Alternatively, an operator with a DVB-H license may be forced to share its network with other providers. For policy makers and regulators it will be important to understand the emerging 'rules of the game' in the electronic communications industry to appreciate what policies may be effective in sustaining the Telecom Reform goals of increased choice, lower prices and increased quality.

Content provider perspective
With the transition to high speed packet access, with data rates into the Mbit/s range, the mobile network has become capable of streaming content to the consumers with adequate quality levels. Hence, the mobile distribution channel has become a contender for the distribution of content to consumers. This means that the availability, quality and functionality offered by the network are becoming important for content providers when deciding on their distribution strategy.

From the value network depicted in Fig. 4 we conclude that the mobile network is one of many ways content can flow from producers to consumers. For the content providers the mobile network appears as 'just another distribution channel'. However, the mobile infrastructure has a unique position in the value network: the mobile network is the only distribution channel for the consumer 'on the move'.

But, alternative channels become accessible as soon as the consumer becomes nomadic or stationary. Moreover, modern mobile devices have a large storage capability; hence, content can be downloaded using the fixed network ahead of time and be viewed while on the move. Hence, real-time needs, convenience and price will be important criteria for consumers in deciding on the use of the mobile channel.

The digital mobile channel is of special interest to provide real time context-aware content. However, despite its potential value early experiments have as yet not led to a 'killer application'. See for instance De Reuver and Haaker [16].

The digital handheld is of special interest as it is becoming the 'preferred device' for consumers. This device can be reached in multiple ways depending on the available connectivity options, and the conscious versus the spontaneous use of the device. The former suggest downloading of the content using the cheapest network in advance; the latter suggests streaming while on the move.

Which distribution channel the consumer chooses is influenced by the mobile provider, but not necessarily prescribed. Creating price parity with alternative communication channels is a strategy that underscores the principle of convenience.

Content providers might be willing to prefer content aggregators using the mobile channel (but not necessarily on a exclusive basis) as mobile subscribers represent a well defined audience.

Mobile communications providers have a direct customer relationship including secure billing. This represents a capability that is of importance to content owners. Nevertheless, content providers may opt for more universal billing arrangements, being transparent to the type of distribution channel being used [17].

The use of the mobile network for content distribution does not directly lead to improved revenues for the mobile provider since there is no carriage fee involved. The mobile communications provider can generate additional revenue by assuming a content aggregator role being paid for by consumers, content providers or advertisers.

Content aggregator perspective

Content aggregators have typically been associated with a specific distribution channel (TV broadcasters, cable broadcasters, satellite broadcaster). Today we find aggregators exploiting the Internet (Apple iTunes, Joost). Aggregators have typically exploited the specific combination of the distribution channel, the device being used, and the way the content is being consumed (e.g., as 'coach potato'). Aggregators are funded through the linkage with the infrastructure (e.g., DVB-C and DVB-S subscription fees) or advertising income (e.g., TV-broadcasting).

The use of the digital mobile channel, the handheld device, and the viewing modalities allow, if not require, a dedicated aggregator role. Given the business model, a mobile channel aggregator will aim at reaching the largest possible number of consumers. Any form of exclusivity will only be of interest if the market size is attractive enough. Aggregation for the mobile channel may be combined with aggregation for the DVB-H channel, depending whether DVB-H substitutes or complements the mobile channel. Typically the aggregator will target high demand segments of the market first, next it will develop a portfolio over multiple segments. Aggregators need to find a source of revenue for a sustainable business model, advertising is one of the options (e.g., Joost). This analysis suggests that aggregators may collaborate with mobile operators but will aim to remain independent.

Mobile communication providers may wish to link up with mobile channel aggregators in providing content to their customers, albeit as said, exclusivity will most likely not be granted. For the mobile provider two options to achieve exclusivity are open: (1) to fulfill the aggregator function through in-house development; (2) to acquire a mobile content aggregator.

To implement the first strategic option mobile providers will need to develop new knowledge and skills, distinct from the skill set available and required for network operations. Parties that have had an aggregator role in the recent past might have the benefit of some knowledge and skills still being available (e.g., KPN-Casema, DT-Kabel Deutschland). Mobile providers will need to find an appropriate target firm to implement the second strategic option.

In the implementation of this strategy the mobile provider may opt to position the service free of charge to its customers, to stimulate network subscription and usage. Positing the service at a premium would place it at a disadvantage compared to similar services generally available over the Internet. To implement the aggregator function the mobile provider will have to resolve the funding issue through advertising, or accept the function as a cost to be offset by growth in subscribers and network usage.

Device manufacturers perspective

For device manufacturers the imperative is to offset market saturation with a device replacement strategy. This calls for devices with new designs, new functionality, and a portfolio of devices allowing more narrow targeting. With the introduction of the iPhone by Apple Inc. the device has become part of an ecosystem that includes content aggregation. The development of DVB-H functionality in mobile devices may be a strategic necessity in the cooperation with mobile communication providers, to avoid a potential deadlock in the development of this technology.

Consumer perspective

In the development of communication and broadcasting the role of the consumer has changed. In a competitive environment the end-user has a wide range of choices and ultimately decides on business success or failure. While mobile communication providers may provide exclusive aggregation services to its customers, it will not be able to offer these exclusive of other aggregators being accessible to its customers. Moreover, consumers may opt for aggregators specializing in terms of content.

5 In Conclusion

Having reviewed the position and perspective of the five major actors in the industry with respect to the delivery of content, two major strategic options are available to the mobile operator faced with technological convergence towards All-IP and a market trend towards content provision.

From an *inside-out* perspective, whereby the resource base of the firm is being leveraged, a mobile operator may aim at becoming the 'distribution channel of choice' for content delivery. This strategic option is in line with the resource based view and assumes a cost leadership role in data transport within the mobile industry, thereby aiming at achieving the largest possible customer base.

In applying an *outside-in* perspective, i.e., in following the market in the development towards content provision, a mobile communications provider may 'move up the value chain' to become a specialized provider in the delivery of content. This option equates to differentiation delivering uniqueness perceived by the customer. This strategic option may be implemented in three steps:

1. Optimising the network for content delivery, i.e. implementing multicast and increasing network capacity,
2. Engaging in aggregation of mobile content,
3. Engaging in shaping content for mobile delivery.

This strategic option requires the firm to either create or acquire new knowledge and skills related to the function of a mobile content aggregator.

In implementing this option we advice the mobile provider to apply a portfolio strategy, whereby the two different sets of resources become part of two different operating units, i.e., not to aim at a tight integration of the two distinct businesses.

As the cost leadership option, aimed at becoming the 'mobile distribution channel of choice', falls within the realm of the current resource base, a successful implementation can be envisioned.

Combining the role of mobile communications provider and mobile content aggregator as described above implies expansion of the resource base of the firm with new assets, skills and capabilities. Whether this strategic option will lead to sustainable competitive differentiation remains to be seen, as the option is subject to economies of scale and scope, and as it implies competition in mobile data transport and in mobile content aggregation.

References

[1] Lemstra, W.: The Internet bubble and the impact on the development path of the telecommunication sector, Dissertation. Delft: TUDelft (2006)

[2] De Wit, B., Meyer, R.: Strategy: process, content, context - An international perspective. Thomson, London (2004)

[3] Van de Kar, E.A.M.: Designing mobile information services Dissertation. Delft: TUDelft (2004)

[4] Bouwman, H., De Vos, H., Haaker, T.: Mobile service innovation and business models. Springer, Berlin (2008)

[5] Bouwman, H., Zhengjia, M., Van der Duin, P., Limonard, S.: A business model for IPTV service: a dynamic framework. Info. 10, 22–38 (2009)

[6] Ballon, P.: Control and value in mobile communications Dissertation. Vrije Universiteit Brussel, Brussels (2009)

[7] Balio, T.: The American film industry. The University of Wisconsin Press, Madison (1985)

[8] Vogel, H.L.: Entertainment industry economics: a guide for financial analysis, 7th edn. Cambridge University Press, Cambridge (2007)

[9] Dunnet, P.: The world television industry: an economic analysis. Routledge, London (1990)

[10] Dake, A., Boers, J.: De kabel: Kafka in de polder. Otto Cramwinckel, Amsterdam (1999)

[11] Abbate, J.: Inventing the internet. MIT Press, Cambridge (1999)

[12] Porter, M.E.: Competitive strategy - Techniques for Analyzing Industries and Competitors. The Free Press, New York (1980)

[13] Braet, O., Ballon, P.: Cooperation models for mobile television in Europe. Telematics and Informatics 25, 216–236 (2008)

[14] Curwen, P., Whalley, J.: Mobile television: technological and regulatory issues. Info. 10, 40–64 (2008)

[15] Rissen, J.-P., Soni, R.: Special issue: 4G Wireless technologies. Bell Labs Technical Journal 13 (2009)

[16] De Reuver, M., Haaker, T.: Designing viable business models for context-aware mobile services. Telematics and Informatics 26, 240–248 (2009)

[17] De Reuver, M., De Koning, T., Bouwman, H., Lemstra, W.: How new billing processes reshape the mobile industry. Info. 11, 78–93 (2009)

Adding Value to the Network:
Exploring the Software as a Service and
Platform as a Service Models for Mobile Operators

Vânia Gonçalves

IBBT-SMIT, Vrije Universiteit Brussel,
Pleinlaan 2, 1050 Brussels, Belgium
vania.goncalves@vub.ac.be

Abstract. The environments of software development and software provision are shifting to Web-based platforms supported by Platform/Software as a Service (PaaS/SaaS) models. This paper will make the case that there is equally an opportunity for mobile operators to identify additional sources of revenue by exposing network functionalities through Web-based service platforms. By elaborating on the concepts, benefits and risks of SaaS and PaaS, several factors that should be taken into consideration in applying these models to the telecom world are delineated.

Keywords: Software as a Service, Platform as a Service, Business Models, Web 2.0.

1 Introduction

The Web has evolved from a large group of informational websites to accommodate a whole new range of services and applications encapsulated in the Web 2.0 concept.

Traditional on-premises software solutions in the domains of enterprise resource planning (ERP), customer relationship management (CRM) and storage are now competing with equivalent Web-based versions. An increasing number of companies like Salesforce, NetSuite or Google are offering their software in a web-based environment providing a much more flexible experience in terms of time and location of access. Over the last decade, this new concept, referred to as Software as a Service (SaaS), has gained attention from Independent Software Vendors (ISVs) and general acceptance by end-users. The benefits of this model over the on-premises model will be explained further in the next section.

Therefore, as software provision in several instances has shifted to the Web, so has the software development cycle. Rather than developing in an offline environment and then testing online, a Web-based application can now be developed, tested, deployed and hosted in a single platform. This new approach, called Platform as a Service (PaaS), enables developers to use a Web-native platform to entirely design and deliver applications in the same environment in which they will run. Both SaaS and PaaS offer a template to monetise data services over the Web.

C. Hesselman and C. Giannelli (Eds.): Mobilware 2009 Workshops, LNICST 12, pp. 13–22, 2009.

Telecommunications networks and the Internet have evolved as disjoint worlds with regards to software applications and application development technologies. Mobile operators are now at the crossroads of maintaining 'dumb pipes' on the one hand, and finding other sources of additional revenue on the other hand. As the IT industry is moving into a Web service delivery model, there seems to exist an opportunity for mobile operators to expose network capabilities and combine these with online content and applications, e.g. mobile service platforms [1]. However, operators' attempts to build platforms for software development so far seem to have failed to attract developers' attention, as can be inferred by the small number of available applications. Consequently, few end-users came into contact with these applications and are benefiting from them. The question is thus whether by building upon SaaS and PaaS concepts and lessons learnt from the IT world, mobile operators could create value for end-users as well as developers, and extract revenues from both sides.

Although both SaaS and PaaS definitions are still open to debate and intertwined with the Cloud Computing concept [2] [3], in this paper both concepts are delineated from a business perspective, taking into consideration the current market trends. Secondly, the benefits and risks of both models are analysed in order to better understand the implications for suppliers as well as customers. Finally, mobile operators' experiences in providing a platform for software development are analysed and several factors that should be taken into account for successfully applying the SaaS and PaaS concepts to the mobile world are outlined.

2   Software as a Service

The Software as a Service (SaaS) concept has been presented by ICT market analysts [4] and software industry associations [5] as an improved version of the Application Service Provider (ASP) model, in which providers host and provide access to a software application over a network. This application service is managed centrally and offered in a one-to-many fashion. The SaaS model has evolved to a web-based application interface, which is significantly different from the ASP model, since this model mostly emulated the traditional client-server philosophy. ISVs were then able to shift from delivering on-premise software solutions to deliver a complete software application to end-users over the Web. SaaS is deemed to incorporate other key attributes such as configurability, multi-tenant efficiency and scalability [6].

Although current literature focuses primarily on business-oriented SaaS services, several examples in the market lead to a categorisation that also includes consumer-oriented service offerings. Therefore, two major categories can be identified. Firstly, *business-oriented services* offered to enterprises and organisations of all sizes, but typically focused on SMEs. The offered services are targeted at facilitating business processes, such as accounting, supply-chain management, and customer relationship management (e.g. NetSuite, Salesforce.com, RightNow). Emergent players offer services that are targeted at niche needs that are usually important for enterprises at a given time or for specific projects, such as storage or computing resources (e.g. GoGrid, Amazon S3 and C2, Mosso, NTRglobal). Secondly, *consumer-oriented services* offered to the general public on a wide range of applications like hosting, news feeds, storage, presentations, and so on. (e.g. Blogger, Flickr, Dropbox, SlideRocket).

The following subsections describe current business models as well as the benefits and risks for the main stakeholders.

2.1 Revenue Models

Shifting from offering on-premise software to SaaS obliged ISVs to change the way software was sold. In an on-premise scenario, the customer buys a license to use an application and installs it on its own or controlled hardware. By owning this license the customer perceives unlimited usage of the software. In the SaaS model, instead of owning a lifetime license, the customer pays a fee for software running on a third-party server and loses access when he ceases payment. This fee can be charged as a pre-paid subscription or on a pay-as-you-go basis. In the examples mentioned above, some players employ both revenue models, but contrasting revenue models can also be found amongst competing players. Typically, in the first approach, players offer different subscription fees based on a combination of allocated resources, namely storage, data transfer, session-time, number of users, etc. In the second approach, pay-per-use fees are charged on the basis of actual usage of those resources. But the granularity of pay-per-use fees can be even greater depending on the target end-user – for instance, Amazon Services applies different pay-per-use prices depending on the geographic location where the data is stored and on the data transfers between different geographical zones. Consumer-oriented services are often provided to consumers at no cost, but are supported by advertising or are offered for strategic reasons such as customer lock-in. These free services are often used to up-sell advanced features (e.g. extra storage space) on a subscription or pay-per-use basis.

2.2 Benefits and Risks

The SaaS concept has clearly changed the way software is developed and delivered to the customer. The shift from on-premises to web applications and from one-to-one to one-to-many service provision has impacted on the relationships between ISVs, customers and developers.

ISVs make and sell software to customers and normally deal with software distribution channels and licensing. In the shift to the SaaS model the most important benefits for ISVs include potential economies of scale in both production and distribution costs, more predictable revenues, development of software with lighter operating system and hardware requirements, and shorter sales cycle. The possibilities of spreading the costs of innovative solutions over many customers, lower license compliance management activities, lower need to deal with piracy and maintain expensive distribution channels can potentially lead to economies of scale in the production and distribution processes. In addition, software with lighter operating systems and hardware requirements enables ISVs to shorten the development life cycle, easily provide additional features that customers ask for, rapid updates and support, and ultimately to potentially reduce maintenance costs.

However, in the SaaS model several risks may arise that ISVs are not used to deal with. Moving to the SaaS model initially requires ISVs to develop new skills when planning SaaS offerings, such as management of billing and customer data, usage metering, security, support services and service provisioning. ISVs should also take up investment in building, and later on, maintaining the IT infrastructure to support SaaS services, or take the approach of partnering with other companies to provide those parts of the system (e.g. DropBox uses Amazon S3 storage services). In this latter case, ISVs now need to manage a network of suppliers that did not exist in the on-premises model. Moreover, with the shift to one-to-many service provision the risk of performance and scalability issues is higher, as many users are using an application running in the same server.

The SaaS concept seems very attractive for business customers such as startups and SMEs which do not see IT-related activities as a core competence and do not have the expertise to develop and maintain applications inside the company. It may enable them to have access to expensive commercial software with little initial investment and low monthly costs, something they could not afford otherwise. Therefore, SaaS offers substantial opportunities to reduce the total cost of ownership of IT resources, alleviating companies from running applications in-house and committing to set-up an IT department to administer software and all related activities involving hardware maintenance, backups, security and user support. For companies with very specific software needs, usually very complex and expensive software used in particular phases of a product development cycle, SaaS can significantly alleviate software costs, as companies would be able to subscribe to services for short periods of time. SaaS providers may also offer customers the possibility to share data with other online or on-premises applications with little effort.

Similarly, the retail customer benefits from access to a wider range of software applications, generally at affordable prices or for free and with little effort of configuration or customization.

For both types of customers, SaaS has the potential to provide a stable, reliable and flexible experience, with access to software regardless of time and location. On the other hand, customers may incur very important risks in exposing or losing business-critical data to platform owners and potentially losing control of data exchanges to third-parties. Likewise, platform owners may lock-in customers by not providing the tools to export (and import) data and therefore making it difficult to switch to other providers. An additional issue significantly different from on-premises software is the interruption of the service due to lack of payment. Consequently, it might also implicate failure to recover stored data.

3 Platform as a Service

The same way the SaaS model is very attractive for companies that cannot afford or do not want to have the burden of investing in hardware and in licensing software, the Platform as a Service (PaaS) model is a novel approach for software suppliers that want to focus primarily on the development cycle and monetisation of new applications, thus bypassing the investment and maintenance of the underlying infrastructure. PaaS platforms facilitate the entire software life cycle by offering the underlying services for application design, development, testing, deployment and hosting [7]. These services enable developers to build on the functionalities of an existing Web platform or SaaS (e.g. MySpace Developer Platform, Force.com) or develop new Web applications (e.g. Google App Engine, Bungee Connect). Although

Amazon Web Services are relevant components for particular stages of the software life cycle, they do not embrace the integral notion of PaaS.

PaaS instantiates the concept of 'Level 3 platform' [8], given the fact that it provides the "runtime environment" to run external applications inside the platform itself. It is likely that PaaS also encompasses Level 1 platform functionality – an access API to allow access to data and services running on the platform – and Level 2 platform functionality – a plug-in API to allow to inject functionality into the platform. For instance, Google App Engine provides a set of APIs to access user-specific data stored in the platform, as well as Google services, such as Calendar events and GMail contacts. In the case of MySpace Developer Platform, applications can run inside or outside the platform, which in the latter case they do via a plug-in API. On the other hand, although Facebook Platform provides the tools for the development and testing of applications, these need to be deployed and hosted outside the platform by the developer, and therefore not constituting a complete PaaS platform according to the given definition.

According to Mitchell [7], a PaaS incorporates six key attributes: integrated environment for development, testing, deployment, hosting and maintenance; delivers a user experience without compromise; built-in scalability, reliability and security; built-in integration with web-services and databases; supports collaboration among developers; and finally, supports deep application instrumentation of application and user activity. Similarly to the previous section, an analysis of the current business models, benefits and risks will follow.

3.1 Revenue Models

Since the software life cycle has many stages, PaaS providers are building their revenue models based on the access to resources in those stages. Some PaaS providers charge a subscription for the access to the development and testing functionalities, while others only charge on the actual time end-users spend interacting with the application, thus after the application is deployed in the market. For instance, the cost for hosting an application with Bungee Connect (currently in beta) is determined by the amount of time users spend interacting with each page of the application. Google App Engine is currently free, but in a future release charging will be based on what the application consumes, up to a budget defined by the application owner [9]. The developer will be able to control daily expenditure by means of adjustable resources, like amount of stored data or incoming and outgoing network bandwidth used.

3.2 Benefits and Risks

In the PaaS model, the most important value element for developers is quite straightforward: develop cloud solutions without having to maintain three environments. In the on-premises software development model, developers work in a development environment, then use a test environment to test it and then move to a third environment for production, which is the only one viewed by the user. This way, a potential faster time-to-market may be expected as developers can test and deploy for production in the same environment. It is also worth highlighting that since the platform owner provides the environments in which the entire software cycle occurs, developers may considerably reduce provisioning and management of their own IT infrastructure (e.g. servers, storage, databases and so on). Currently, developers may try out different PaaS platforms at no cost for the developing and testing phases, enabling them to assess the functionalities of that platform and whether it facilitates integration with other online services. In this case, developers may consider as main costs the runtime fees based on the actual computing resources used by the applications or the number of users using them. Some PaaS platforms already enclose the same feeling of a developer community, by giving developers the means to communicate, share knowledge and reuse other developers' code.

PaaS platforms can be more or less open and, in the latter case, lock-in developers. Some PaaS platforms are based on standard programming languages like WyaWorks (based on Java), on a combination of standard programming languages and proprietary add-ons like Google App Engine (based on Python), or on proprietary programming languages like Force.com (based on Apex). Therefore, developers may be confronted with a long learning curve to master proprietary programming languages and APIs, thus putting more effort in developing an application in comparison with a development environment of an open platform. A closed PaaS platform may create lock-in by preventing developers to port an application to another platform, which may only be circumvented by building the application from scratch in the new platform. Closed platforms may also prevent developers from integrating data from third-party platforms or services to create, for instance, mashups and therefore hinder innovation. The whole concept of providing developers with all development tools may also create a fundamental gap with traditional development environments – the developer is limited to use the APIs available in the platform, preventing him to give wings to new ideas by extending or incorporating new functionalities in APIs.

Like in SaaS models, platform owners can make platforms available with light operating systems and hardware requirements, both for the development phase and for the runtime. Similarly, the platform owner would avoid dealing with license compliance and piracy issues. Still, the most important value element for platform owners is earning revenues from applications hosted in the platform, which might well be the next killer applications.

Nevertheless, the success of the platform will heavily depend on the platform owner's strategy and the right balance between open and closed, proprietary and non-proprietary, free and paid, and so on, making it captivating enough for developers. It is likely that developers are not keen on relearning everything, that they want to try out how the platform works and responds, are typically eager to use tools to share ideas and knowledge, and ultimately want the option to put the application on the market with little effort. But once the platform is adopted, a platform owner has to deal with a whole set of issues related to performance, scalability, storage and security. The platform owner cannot control the quality and security of the code that will run within the platform, but has to create the technical mechanisms to anticipate the consequences of such problems. Additionally, many of the analysed platforms detail in a terms of service agreement the possible consequences in case of overuse of computing resources or faulty applications that disrupt the platform.

One can argue that in this model the developer can devote more time to the creative process and therefore a whole new range of innovative applications will emerge. The end-user will benefit from applications that fully realise the Web 2.0 principles making more use of integration and aggregation of data provided by other applications or systems. The enormous offer of this type of applications will likely decrease the price for consumers. The end-user might however incur the risk of paying for a service that is periodically unavailable. Considering Google App Engine's business model, developers are able to set quotas for computing resources in order to prevent the cost of the application from exceeding developer's budget [9]. When the application exceeds the allocated resources, App Engine will return an unavailable status and the user will not be able to use the application.

4 Summary of Benefits and Risks

Although SaaS and PaaS might target different customers – SaaS is focused primarily on business and retail customers and PaaS targets customers and developers – they appear to be based on the same revenue models of subscription or pay-per-use fee, and have some common benefits and risks. The diagram of Fig. 1 summarises the benefits and risks for the three aforementioned stakeholders – platform owner, end-user and developer. The overlapping benefits and risks of the two models are represented in the central section of the diagram.

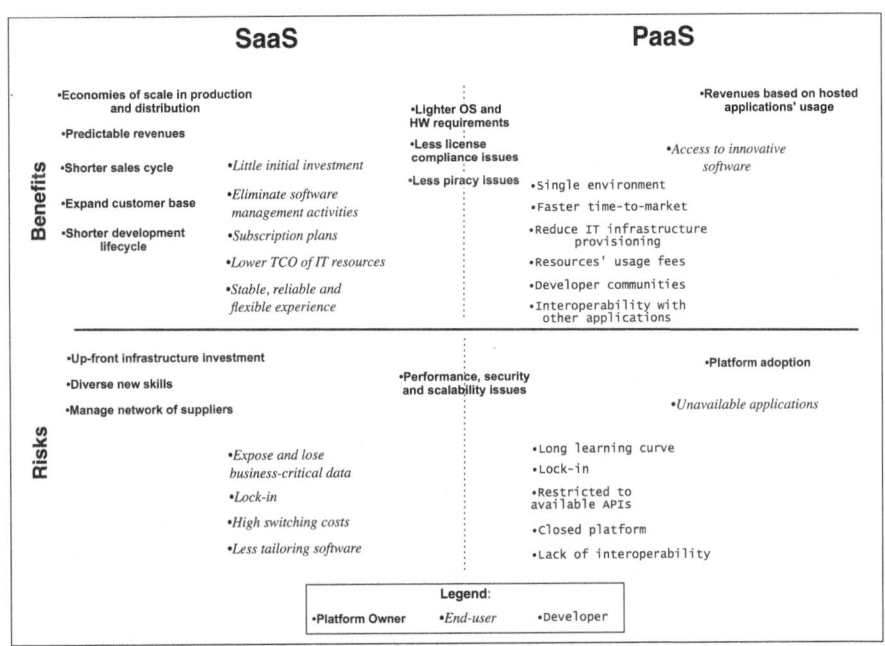

Fig. 1. Summary of benefits and risks for different stakeholders in Software as a Service and Platform as a Service models

5 Analysis of Emerging Experiences of Mobile Operators' Service Platforms

This section analyses the emerging experiences of mobile operators in building platforms for software development targeting end-users and developers. By applying the benefits and risks of SaaS and PaaS concepts to the mobile world, potential drivers for platform adoption are identified.

In case a mobile operator chooses to develop in-house applications targeting retail or business consumers, the SaaS model is applicable. Therefore, a mobile operator would be required to build a team of developers to create and support applications, but it seems rather unlikely that an operator would be willing to take up this task. Mobile operators have always been focused on network operations and few attempts to diversify services through software service provision have been made in the past.

One may argue that mobile operators' thematic portals offering ringtone or game downloads are an instance of a SaaS platform. However, for these portals operators mainly partnered with content and application providers. Moreover, in general these portals did not sustainably attract consumers because they failed to provide a flexible and rich experience and engaging services. For these reasons, a PaaS scenario in which operators rely on third-parties to develop applications seems to be the opportunity to expose telecommunications network functionalities, as long claimed by developers. Some major operators already launched platforms that incorporate the PaaS concept – Ribbit, incorporates former Web21C SDK (BT), Orange Partner (Orange), Litmus (O2), Betavine (Vodafone) and open movilforum (Telefonica).

Blending the Web 2.0 development environment with the network's services and information would result in a richer application environment with personalised services. Mobile operators hold plenty of information about their customers and a large customer base. That means an application could be passing code to be executed on the operator's side and send the resulting information to the end-user, but still keeping user's information private. Thus revenue opportunities for both suppliers and operators are significant.

In order to allow this integration of both worlds, mobile operators need to provide application programming interfaces (APIs) that give the means to developers to efficiently explore their networks' capabilities such as location, rich presence, charging, messaging and so forth. Perry [10] and Warfield [11] analysed current market PaaS and identified several advantages in providing a common and open API. A common API would stimulate innovation, reduce barriers to entry and increase portability between operators. Additionally, it would reach out to more developers, as these would easily identify the advantages of coding once and deploy and monetise in several platforms [12]. Current operator platform's initiatives are using distinct APIs to give access to theirs networks – GSMA OneAPI [13] and OMA IMS [14].

A single environment for the entire application life cycle and the potential of faster market delivery are key advantages of a PaaS platform for developers. Still, a strong developer community around a platform gives developers a feeling of control over the platform and might also be crucial to guarantee platform adoption. More developers means still more developers, which means more applications, which means more customers, and so on. Quayle [15] studied forty application developer communities and identified six important factors in building a developer community, among which is included a channel to enable developers to earn profits from their applications. In a mobile telecom context, a marketplace would allow a win-win situation between mobile operators that sell and host the applications, developers that write them and customers that want to buy and use them. Additionally, an operator would need to promote this marketplace as it own service, by giving it visibility through the operator's website and marketing campaigns.

Regarding current operator platforms all have succeeded in building up a community providing code examples, forum, blog, and case studies. Almost all provide optional hosting of the application, except for Orange Partner that partnered with Cellmania. Ribbit, Orange Partner and Litmus already gave a commercial component to the platform, enabling the developer to define a price and sell an application; still, only Orange succeeded in incorporating applications into the operator's website via the Application Shop section [16]. Litmus, on the contrary, transmits the idea that the whole concept is still a trial by incentivising end-users to test applications in return for a free application. Ribbit charges developers at production stage based on the resources the application uses. In Orange Partner, the revenue is split between Orange, Cellmania and the developer, but it is not clear whether Cellmania charges developers on a resources usage basis.

To conclude, a PaaS platform brings mobile operators the opportunity to expose their networks to the Web 2.0 developer community. The operator will play a role of intermediary for the distribution of third-party applications to its customer base, earning revenues from both sides. For this scenario to be possible, it is necessary to engage both developers and customers by building developer communities and presenting developers a direct route to the market through a marketplace that effectively reaches the customer. A common API will further engage developers and leverage the potential of reaching customers across operators and countries, just like a regular Web-based application would. However, up to now mobile operators' platforms have been slow to mature and offered APIs lack consistency.

6 Conclusion

This paper analysed the SaaS and PaaS concepts from a business perspective and highlighted the benefits and risks for both suppliers and customers. It then explored the potential of applying these models to the mobile world to allow for additional sources of revenue. The PaaS model seems to best suit a scenario in which an operator exposes telecommunications network functionalities by creating the platform for third-parties to develop applications that offer a flexible and rich experience and engaging services, and acts as an intermediary in the distribution of these applications to its customer base, earning revenues from both sides. Three main factors were identified as drivers for platform adoption: a common API, a developer community and a marketplace. Current mobile operators' platforms have been slow to integrate a marketplace and offered APIs are fragmented among operators.

Acknowledgement

This work was performed within the WTEPlus project (number 10328) funded by the Interdisciplinary Institute for Broadband Technology (IBBT). The author gratefully acknowledges Dr Pieter Ballon for helpful comments on earlier versions of this paper.

References

1. Ballon, P.: Control and Value in Mobile Communications: A Political Economy of the Reconfiguration of Business Models in the European Mobile Industry: SSRN (2009)
2. Armbrust, M., et al.: Above the Clouds: A Berkeley View of Cloud Computing, EECS Department. University of California, Berkeley (2009)
3. Chappell, D.: A Short Introduction to Cloud Platforms. DavidChappell & Associates (2008)
4. Mizoras, A., Goepfert, J.: AppSourcing Taxonomy and Research Guide. In: IDC (ed.) IDC-Industry Developments and Models (2003)
5. SIIA (ed.): SIIA: Software as a Service: Strategic Backgrounder, S.I.I. Association, Washington (2001)
6. Chong, F., Carraro, G.: Building Distributed Applications: Architecture Strategies for Catching the Long Tail (2006),
   http://msdn.microsoft.com/en-us/library/aa479069.aspx
7. Mitchell, D.: Defining Platform-As-A-Service, or PaaS (2008),
   http://blogs.bungeeconnect.com/2008/02/18/
   defining-platform-as-a-service-or-paas/
8. Andreessen, M.: The three kinds of platforms you meet on the Internet (2007),
   http://blog.pmarca.com/2007/09/the-three-kinds.html
9. Google App. Engine's Quotas (2009), http://code.google.com/appengine/
   docs/quotas.html#Adjustable_and_Fixed_Quotas
10. Perry, G.: Thoughts on Platform-as-a-Service (2008),
    http://gevaperry.typepad.com/main/2008/09/
    the-future-of-p.html
11. Warfield, B.: The Perils of Platform as a Service (It's Not As Bad As All That!) (2007),
    http://smoothspan.wordpress.com/2007/10/08/the-perils-of-
    platform-as-a-service-its-not-as-bad-as-all-that/
12. Orange sees API uptake, but operator cooperation needed (2009),
    http://www.telecoms.com/itmgcontent/tcoms/features/articles/
    20017613596.html
13. GSMA: 3rd Party Access Project,
    http://gsma.securespsite.com/access/default.aspx
14. Alliance, O.M.: OMA IP Multimedia Subsystem (IMS in OMA) V1.0 (2005)
15. Quayle, A.: Opening Up the Soft Service Provider: The Telco API (2008)
16. Orange Launches New Widget Experience and Expands Multi-Platform Application Shop (2009),
    http://mobileworldcongress.mediaroom.com/
    index.php?s=43&item=525

Business Model Evaluation for an Advanced Multimedia Service Portfolio

Paolo Pisciella[1], Josip Zoric[1,2], and Alexei A. Gaivoronski[1]

[1] Faculty of Social Sciences and Technology Management,
Norwegian University of Science and Technology, Trondheim, Norway
[2] Telenor R&D, Trondheim, Norway
{paolo.pisciella,alexei.gaivoronski}@iot.ntnu.no,
josip.zoric@telenor.com

Abstract. In this paper we analyze quantitatively a business model for the collaborative provision of an advanced mobile data service portfolio composed of three multimedia services: Video on Demand, Internet Protocol Television and User Generated Content. We provide a description of the provision system considering the relation occurring between tecnical aspects and business aspects for each agent providing the basic multimedia service. Such a techno-business analysis is then projected into a mathematical model dealing with the problem of the definition of incentives between the different agents involved in a collaborative service provision. Through the implementation of this model we aim at shaping the behaviour of each of the contributing agents modifying the level of profitability that the Service Portfolio yields to each of them.

Keywords: Service Platforms, Business Models, Multi Follower, Stochastic Programming, Video on Demand, IPTV, User Generated Content.

1 Introduction

Advanced data services are customizable and oriented towards the on-demand and real time delivery. Typically they are bundled in Service Portfolios. Thus it is important to develop and provide such services in a flexible way. Services Portfolios are composed of different enablers and basic services provided by a constellation of actors. Most of the reseach carried out so far has been oriented towards the technical aspects of such a service provision, while the quantitative evaluation of business modeling aspects is still in an early phase.

The aim of this paper is to provide a proof of the concept for the economic evaluation of collaborative provision of a data service focused, for the time being, on the aspects of economic viability and risk management. We will focus on the case of the provision of a service composed by the bundle of three different services given by Internet TV (IPTV), Video on Demand (VoD) and User Generated Content (UGC). The composed service is used in a particular business scene, such as a concert.

We will analyze how the architecture of such a service delivery system is composed, in order to emphasize the elements which are technically relevant

C. Hesselman and C. Giannelli (Eds.): Mobilware 2009 Workshops, LNICST 12, pp. 23–32, 2009.
© ICST Institute for Computer Sciences, Social-Informatics and Telecommunications Engineering 2009

w.r.t. a business model. At the same time, considerations about the economic values will be introduced in order to consider a quantitative revenue and risk management model. Then the model is applied for fine tuning, via the use of a revenue sharing scheme, the decisions of the service providers about the amount of service to supply to the service portfolio to be bundled. Such model, has been tested in earlier work [1] for different configurations of the sharing scheme resulting in a feasible service portfolio provision.

In this paper we describe an implementation to select of the optimal sharing scheme in order to make the service portfolio feasible and to deliver the highest possible return on costs to the agent assuming the role of aggregator. The approach is based on an *aggregator centric business model*, described in [8], where an actor takes the double role of Service Aggregator, offering the service portfolio to the end user, and Portal Provider that visualizes the contents to the end user. In our approach we assume this actor to cover the further role of Platform Operator, offering the tools necessary for bundling the service portfolio. The analysis considers a general notion of platform, intended as an entity coordinating interactions between two or more distinct groups of stakeholders [8].

The rest of the paper is organized as follows. Section 2 describes three user scenarios where the service portfolios are invoked by the end user. The technical analysis of the provision of each basic service bundling the service portfolio is done in section 3 to understand the technical bottlenecks of such a provision. Section 4 describes how the technical constraints, together with business considerations, are mapped in a quantitative model related to the selection of a revenue sharing scheme amongst the service providers. Section 5 discusses the results of the implementation of the mathematical model. We conclude the paper and outline the future work in section 6.

2   Typical Business Scenes

The purpose of this section is to describe three user scenarios, in which the end user invokes the use of a Service Portfolio composed of the three multimedia services discussed above, in order to position our problem within a business point of view.

The first scenario depicts the end user at a concert. Using her device, the end user can have the concert broadcasted on enhanced TV while, at the same time, the end user can use the VoD service to retrieve material about the band and access to UGC directly on the device. In the second scenario the end user is a tourist using video guides requested as VoD together with UGC in order to read comments about places to visit, during the trips she may eventually want to relax watching IPTV or some desired VoD. A third service allows the users to interact with TV requesting VoD for changing the storyline of a movie, involving use of forums, chats and other UGC.

These services are provided collaboratively by a set of Service Providers that can decide to supply their component to a set of Service Portfolios. Their goal is to select an appropriate group of Service Portfolios that sounds as most profitable.

We consider the VoD provider to take the role of Platform Operator. Her work is to ensure that each Service Provider supplies a minimum amount of Megabits per second (Mbps). Such a coordination is achieved by a wise choice of the revenue sharing scheme between the contributing actors.

3   Technical Analysis of the Service Portfolio Provision

We assume that all the basic services are delivered by exploiting a high-bandwidth backbone network such as ATM or SONET [4]. This will allow us to neglect bottlenecks at the network level (in this case physical layer) and concentrate on possible constraints in the application layer of the system.

3.1   System Architecture Design

Video on Demand

In the general architecture of a VoD system we have thousands of local distribution networks, delivering audio and video content stored in Video Servers, connected by a high-bandwidth backbone network.

Video Servers are capable of storing and outputting a large number of movies simultaneously. Such a storage is done using different kind of supports characterized by a tradeoff between cost and number of users that can be served. A Video Server is composed by one or more high performance CPUs, a shared memory, a massive RAM cache for content which is supposed to be requested with more frequency, a variety of storage devices for holding the movies and some network hardware such as an optical interface (call it for example Network Interface) to a SONET or ATM backbone. These subsystems are connected by an extremely high speed bus (at least 1Gb/sec.). The CPUs are used for accepting user requests, locating movies, moving data between devices. Many of these operations are time critical, and thus they constitute a constraint for the number of services that can be produced in a given frame time.

We will assume that the VoD provider has constraints with respect to the number of services feasible for a given time unit, and such a constraint depends on the capacity of the server. The performance of the system is measured in Megabits per second (Mbps).

Internet Protocol Television

IPTV uses IP networks to deliver digital TV programming from a central location to a base of multiple subscribers. A large IPTV network works in the follwing way (see, for further discussion [6]). The Super-video Head End (SHE) receives video signals for national programming from a variety of sources and in different formats and encryption protocols. Its task consist in decripting the source format, converting it in a suitable data type and rebroadcasting it to regional Video Head ends Offices (VHO). When the VHO receives the packets, it aggregates such a content with the local content encoded and forwards it to the

Video Switching Office (VSO) which is the phone company's infrastructure for delivering video to users. In order to deliver IPTV to the end-user the carriers plan to use IP multicasting. With multicasting, a single copy of a program is transmitted through the network and replicated only where necessary. One of phone carriers' biggest IPTV challenges will be having enough bandwidth for both the potential growth in data-intensive high-definition programming and a growing number of customers (see e.g. [5]). To model the technical process necessary to deliver the IPTV service we will consider in a deeper detail the operations performed by each of the elements constituting the hardware architecture of an IPTV system.

User Generated Content

User Generated Content (UGC) refers to various kinds of media content, publicly available, that are produced and shared by end-users, whether is it a comment left on Amazon.com, a professional quality video uploaded to YouTube, or a student's profile on Facebook [7]. For capacity constraint purposes we will consider only the aspects of UGC including multimedia content, which is characterized by bottlenecks at the service provider side. We will approximate the architecture of a UGC system for multimedia content by considering it an extension of the VoD case, but allowing for the upload of multimedia content by the end-user.

3.2   System Architecture Approximation

The architecture of a system as the one described above is very complex and the elements involved in the provision of the services need to perform a large amount of operations. However, not all the operations are considered as critical for the provision of a Service Portfolio, since modern technology can efficiently perform most of them. For this reason we have restricted the analysis only to the operations that constitute a bottleneck in terms of capability to serve a continuously increasing number of end users. The approximation of the overall architecture follows the guidelines discussed in [2] and [3], and can be formalized through a UML class diagram, like the one showed in figure 1. The first element considered in the figure is a goal, which is a state a user tries to achieve or obtain. What is delivered to the end user is a Service Portfolio, i.e. a collection of services. Enabler is a functionality used by a service provider to create a service. Such enablers provide the SP capabilities and consume various resources, which constitute a constraint on the maximum level of basic service that can be provided to the Service Portfolio. Class diagrams provide an approximation of the system architecture for the different providers and highlight the importance of resources consumption in the service delivery process. To deliver a given service

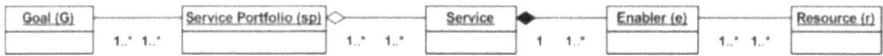

Fig. 1. A Generic Service Platform Model (source: Zoric, 2008)

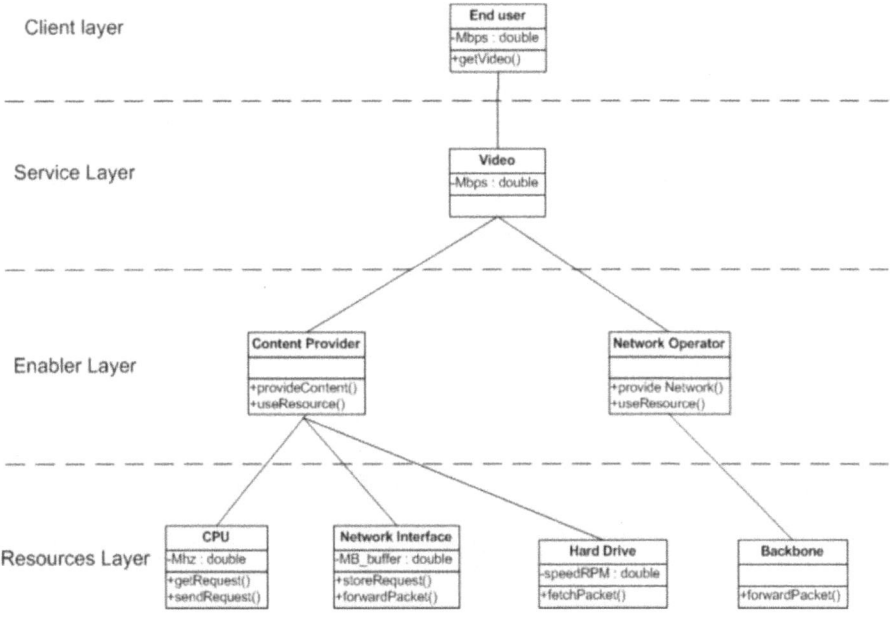

Fig. 2. Class diagram for the VoD case

(e.g. a video or some multimedia content) the server has to be able to provide a minimum amount of Mbps, which are a function of the number of operations that the server can perform.

Video on Demand

In figure 2 we identify the relations between the classes of objects describing the VoD system elements in a four layer architecture: Client Layer, Service Layer, Enabler Layer and Resources Layer. The Client Layer contains the class of objects sending the request for a service. The end user is characterized by the attribute defining the number of Mbps requested. At the Service Layer we have the video, which is characterized by a the actual number Mbps. The set of functionalities used to create the services, considered at the Enabler Layer is given by two enablers: content provider and network operator. These enablers provide the capabilities to build up a service ensuring a minimum amount of Mbps to each end user, in order to achieve a satisfactory experience. At the same time enablers consume resources, which are described in the Resources Layer. Let us examine with more detail the considerations introduced about the mapping between operations performed by the elements of the server and the Mbps delivered. We will consider, for the case of VoD, a table with four resource-consuming operations. The same considerations are applicable to IPTV and UGC cases. Namely, the service produced by the VoD provider is the result of the following main operations:

Operation	Performed by	Bounding measure
accepting download requests	CPU	Mhz
requesting data	CPU	Mhz
fetching data	Hard Drive	rpm
delivering data	Network Interface	MB

Given the power of each of these elements, there is an upper bound on the number of operations that they can perform in a particular frametime, bounding in the same way the number of Mbps that can be delivered.

Internet Protocol Television
The analysis considers the same four layers as previously done. In the Client Layer we have a group of end users sending a connection request to a set of different TV channels, which in our analysis are positioned in the Service Layer. The enablers involved in this delivery are the Content Provider (for the encoding and forwarding processes) and the Context Provider (for what concerns the context detection for local and customized content) on one side and the Network Operator (for the physical delivery of data) on the other side. What is done at the server side now is not so different from the Video on Demand case, except for the fact that the CPU has to handle the conversion of the source data obtained from an external antenna or satellite (receiving, decoding and re-encoding in a suitable format). The VHO performs the same operations as the SHE (receive and convert source), but it operates with local content. Moreover, it bundles the local content with the global one. The VSO has the task of demultiplexing the packets transporting data of different TV channels to different end users.

User Generated Content
As mentioned earlier, the UGC architecture resembles the one for the VoD, with the possibility for the end user to upload multimedia content to the server. In order to approximate the architecture of such a system we just take as a reference what is showed in figure 2, changing the class *Video* with the class *Content* and adding the operation loadContent() in the end user class and the operation storePacket() in the Hard Drive class. The operation getRequest() in the CPU class can be thought as well as a download request or an upload request. In the same way the operation forwardPacket() in the Network Interface class can be thought as a duplex connection: packets flow in both ways.

4   Modelling the Collaborative Service Provision

In this section we consider the model introduced in [1], which aims at achieving a collaborative provision of a data service via the choice of a revenue sharing scheme between the providers, taking as input statistcs on profitability of the service as well as preferences of the component providers in terms of risk tolerance and minimum accepted level of profitability.

The main building blocks of the model are *services* given by VoD, IPTV and UGC and indexed by $i = 1 : 3$ and *Service Portfolios* defined by the collection of services provided to the end user in the concert scenario, the tourist scenario and the interaction scenario which are indexed by $j = 1 : 3$. The relation between services and Service Portfolios is described by the coefficients λ_{ij} which measure the amount of Mbps that service provider i has to supply for delivery of the unit amount of Service Portfolio j. Thus, a Service Portfolio j can be described by vector

$$\lambda_j = (\lambda_{1j}, .., \lambda_{3j}) \tag{1}$$

which describes the amount of Mbps each service provider has to deliver to provide one unit of service portfolio j.

An end user requests a service portfolio j and generates a revenue v_j per unit sold. We can think v_j as a random variable; this means that the price level to be set in order to keep a given level of demand is uncertain. Randomness is thus due to uncertainty of user acceptance and demand.

A way to quantify the uncertainty attached to the revenue of each service portfolio is given by the computation of the risk level associated. In this framework we will consider risk as the non predictability of the unit revenues in order to have a certain level of demand, and this allows us to use standard deviation, one of the most established and wide used risk measures in finance, as a risk measure for the unit revenue. The economical concept underlying such a kind of analysis is that Component Providers want to choose how many Mbps to supply to each service portfolio in order to get the highest return on costs possible for a given level of risk. Increasing the level of risk accepted will result in a higher level of expected return on costs. We assume that the revenue obtained by a component provider is a share of the revenue generated by the sale of the Service Portfolio. Revenue is proportional to the percentage of Mbps provided to such bundle with respect to the amount of Mbps necessary to supply a enjoyable service to the end user.

The revenue v_j generated by a unit of Service Portfolio j is distributed among the actors who participate in the creation of the service portfolio using a vector of revenue shares

$$\gamma_j = (\gamma_{1j}, .., \gamma_{3j})$$

Determination of these revenue sharing coefficients is one of the objectives of the design of the business model for service provision.

Let us consider c_i as cost per unit of Megabit of service i sent per second and x_{ij} as the portion of provision capability (Mbps sent over total Mbps that the provider can supply) for service i dedicated to participation in provision of Service Portfolio j. We can define the expected return on total costs of the i-th provider as

$$\bar{r}_i(x_i, \gamma_j) = \sum_{j=1}^{4} \mu_{ij} x_{ij} = \sum_{j=1}^{3} x_{ij} \left(\gamma_{ij} E \frac{v_j}{c_i \lambda_{ij}} - 1 \right) + x_{i4} \left(E \frac{v_{i4}}{c_i} - 1 \right) \tag{2}$$

at the same way, we can define the standard deviation of the return on costs as

$$R(x_i) = \text{StDev}(r_i(x_i)) = \text{StDev}\left(\sum_{j=1}^{4} r_{ij} x_{ij}\right) \tag{3}$$

The problem of the i-th Service Provider is given by finding the best distribution of her service, in trems of Mbps supplied, amongst the Service Portfolios under a risk-performance point of view. Following the framework provided by [1] we define the problem of the i-th Service Provider as

$$\max_{x_i} \bar{r}_i(x_i, \gamma_j) \tag{4}$$

subject to constraints

$$\sum_{j=1}^{4} x_{ij} = 1, \ x_{ij} \geq 0 \tag{5}$$

$$R(x_i, \gamma_j) \leq \bar{R} \tag{6}$$

where we emphasize here the dependence of risk and return on the revenue sharing scheme γ_j. To provide a satisfactory experience to the end user, every component provider involved in the creation of the service portfolio has to provide a level of Mbps larger or equal to a given treshold. The solution of such a portfolio problem will be denoted by $x_i(\gamma_j)$ for all generic actors providing service i for the Services Portfolio j.

We assume that one of the Service Providers covers a further role of Service Aggregator. It is up to her to decide the correct revenue sharing scheme in order to compose the Service Portfolio. Formally such a provider faces the following optimization problem:

$$\max_{\gamma_j} \bar{r}_1(x_1(\gamma_j), \gamma_j) \tag{7}$$

subject to constraints

$$x_{ij}(\gamma_j) \geq x_{ij}{}^{\min} \text{ for all } i \in I_j \tag{8}$$

$$\gamma_j \in \Gamma_j \tag{9}$$

where the set Γ_j can be defined, for example, by

$$\left\{\gamma_j | \gamma_{ij} \in [0,1], \sum_i \gamma_{ij} = 1\right\} \cap \left\{\gamma_j | p_i^* - \Delta^- \leq \frac{\gamma_{ij}}{\lambda_{ij}} E v_j \leq p_i^* + \Delta^+\right\}$$

where p_i^* is a target for the price of the service i and Δ^+ and Δ^- defines the tolerances within which she is willing to accept a different price, and

$$I_j = \{i : \lambda_{ij} > 0\}$$

The problem defined in (4)-(9) presents the structure of a stochastic bilevel multifollower model.

5   Numerical Implementation

The previous model has been tested in [1] computing the value of the objective function (7) corresponding to different values of the sharing scheme chosen for a platform service portfolio as shown in the following table.

Table 1. Values of the optimal expected return on costs of the Platform Operator (also VoD provider) as a function of the sharing scheme weights granted to the IPTV provider (first column) and the UGC provider (first row)

0	0,474	0,4795	0,485	0,4905	0,496	0,5015	0,507	0,5125	0,518
0,1765	0	0	0	0	0	0	0	0	0
0,182	0	0,2312	0,2239	0,2163	0,2083	0,2001	0,1917	0,1834	0,1760
0,1875	0	0,2239	0,2163	0,2083	0,2001	0,1917	0,1834	0,1760	0
0,193	0	0,2163	0,2083	0,2001	0,1917	0,1834	0,1760	0	0
0,1985	0	0,2083	0,2001	0,1917	0,1834	0,1760	0	0	0
0,204	0	0,2001	0,1917	0,1834	0,1760	0	0	0	0
0,2095	0	0,1917	0,1834	0,1760	0	0	0	0	0
0,215	0	0,1834	0,1760	0	0	0	0	0	0
0,2205	0	0,176	0	0	0	0	0	0	0

The model (4)-(9) has been successively implemented using the commercial software MATLAB with the aim of finding an optimal solution. We used anonymous data units in order to keep it similar to the original problem test. To find the optimal solution of the model considered at the beginning of the section we defined a function which takes as input the data necessary to run a set of mean-variance portfolio optimizations. In particular the input corresponds to the variance covariance matrix for the Portfolio Services revenues, the prices of the Portfolio Services, the costs for transmitting one Mbps of service as well as the risk bounds and the minimum level of basic service to provide to the service bundle in order to make the provision feasible. The output provided by the function is the amount of service supplied by each service provider to the service portfolio for a given sharing scheme. Such a function is used as constraint for the upper level optimization which attempts to find the optimal sharing scheme to provide the highest returns to the platform operator.

The result of the optimization gives an optimal sharing scheme vector of 34,15% to the VoD provider/Platform Operator 17,98% to the IPTV provider and 47,87% to the UGC provider, with both the IPTV provider and the UGC provider having expected returns at their minimum accepted level and risk measure value at their maximum accepted value. The optimal expected return of the Platform Operator (VoD provider) correspondent to the maximum risk level accepted by this provider is given by 23,51%.

6   Conclusions and Future Work

We have discussed the implementation of a model designed to coordinate the provision of services to bundle a Service Portfolio through the choice of a suitable revenue sharing scheme. The solution ensures that each service provider participates actively in the creation of a customized Service Portfolio and, at the same time, allows the Platform Operator to obtain the highest possible return. In the business model evaluated one agent takes the role of coordinator between the different providers, while other Platform Business Models (see e.g. [8] or [9]) with different coordination rules require different evaluation methods. The implementation considers Mbps as the relevant measure for the determination of the service provision level and it seems to be quite suited for multimedia contents, but such a measure could not be used in case of service providers not having bottlenecks linked to bandwidth.

Thus, the model is yet considered at an early stage and not fully operating in an industrial basis. A possible future extension of the evaluation framework can proceed in the direction of including different types of contribution models, in order not to restrict the analysis to multimedia bandwidth critical content. The model considers as well a set of parameters not always easy to estimate from market data, a way to simplify the constraints in order to allow for an easier evaluation of the parameters needed is considered a right step in the path of merging the academic and the industrial perspectives.

References

1. Gaivoronski, A.A., Zoric, J.: Business models for collaborative provision of advanced mobile data services: portfolio theory approach. Operations Research/Computer Science Interfaces Series, vol. 44, pp. 356–383. Springer, Heidelberg (2008)
2. Zoric, J.: Practical quantitative approach for estimating business contribution of enablers and service platforms. In: Proceedings of ICIN 2008 - Services, Enablers and Architectures supporting business models for a new open world, Bordeaux, France (2008)
3. Zoric, J., Strasunskas, D.: Techno-Business Assessment of Services and Service Platforms: Quantitative, Scenario-Based Analysis. ICT-Mobile Summit (2008)
4. Tanenbaum, A.: Computer Networks. Prentice Hall, Inc., Englewood Cliffs (2004)
5. Ortiz, S.: Phone companies get into the TV business. IEEE computer society (2006)
6. Cisco Wireline Video/IPTV Solution Design and Implementation Guide (2004)
7. IAB Platform Status Report. User Generated Content, Social Media and Advertising – An Overview. In: Interactive Advertising Boureau (2008)
8. Ballon, P., Walravens, N.: Competing Platform Models for Mobile Service Delivery: the Importance of Gatekeeper Roles. In: 7th International Conference on Mobile Business (ICMB 2008), Barcelona (2008)
9. Strasunskas, D. (ed.): Revised business analysis and models. SPICE project deliverable D1.7, IST-027617 (2008)

Connecting Business Models with Service Platform Designs – Quantitative, Scenario-Based Framework

Josip Zoric

Telenor R&I, and Norwegian University of Science and Technology,
Trondheim, Norway
josip.zoric@telenor.com

Abstract. Heterogeneity of technical and business designs, complexity of collaborations, and incentives are just some of the consequences of service platform evolution that complicate their business analysis. Business models are usually on a higher abstraction level than service platform designs, which requires detailing prior to their financial analysis. This work proposes a framework for quantitative analysis, which "reinterprets" the business models by underlying service platforms technical and business entities, processes and scenarios. In such a way it prepares them for business analysis and valuation, focusing also on incentives of collaborating business actors. We explain the approach theoretically and demonstrate its use on the proof-of-the-concept service platform.

Keywords: techno-business modeling, service platforms, scenario approach, service, enabler, quantitative analysis, valuation.

1 Introduction

Converged, cross-media platform models and mobile communication mashups and platforms are increasing opportunities for new – and more dynamic value propositions. Heterogeneity of technical and business designs, complexity of collaborations, diversity of roles and incentives are just some of the consequences of service platform evolutions, which make their techno-business analysis complex. Service platforms are complex technical and business systems, hosting multiple service portfolios (bundles), influenced by dynamics from four spheres: user, business, system and technological [1-3]. All of them can dramatically influence the quality of experience, quality of service and thus the business value. A good quantitative business estimate is dependent on the input of all four spheres. The business analyses should not oversimplify any of them.

Technical and business models of services and platforms revolve around two knowledge sets:

(1) *At a higher abstraction level* - business models (BM) and qualitative business analyses specifying the value propositions and underlying value configurations (e.g. [4-8]). There is neither consensus nor a standard approach in business

C. Hesselman and C. Giannelli (Eds.): Mobilware 2009 Workshops, LNICST 12, pp. 33–44, 2009.
© ICST Institute for Computer Sciences, Social-Informatics and Telecommunications Engineering 2009

modeling [4-8]. Pateli and Osterwalder [14,6] suggested: whoever wishes to build a BM should select one of several published approaches, justify why using that instead of another similar, and use that as foundation for building her/his model.

(2) *At a higher detail level* – a plethora of technical and business designs of services and platforms (difficult to analyze and compare).

But how to map the BMs to their realizations, and how to estimate their technical and business impacts. In business terms, how to valuate the underlying asset (in our case service platforms) and quantitatively analyze business actor incentives in collaborative service provision? We offer a framework for techno-business analysis (TBA) of services and platforms, which might serve this purpose. The TBA [1,2] engages in service and SP analysis with respect to four important aspects affecting its technical and business performance: user, business, system and technical, which we discuss below.

User analysis in our approach contains two aspects: service customer and service user. Customer aspect focuses on user value concepts, their business counterparts and contribution of services in their realization. Focusing on service user means mapping the user value concepts to goals, requirements and usage scenarios, and analyzing which service can support their realization. We call it a service usage modeling. We focus on service support in delivering the value to the user (assisting the user in goal realization that is represented by scenarios). A user scenario describes the end-user behavior when interacting with the service.

Business analysis considers value proposition, value configuration, corresponding business models [4-8], and business process models [2] specifying collaborations in service delivery.

System and technical analyses focus on system models and system process models needed for analyzing system interactions in collaborative service provision. In this part of the TBA we take over the business process models and map them to their system and technical counterparts, with help of scenario techniques. We use (Message Sequence Charts (MSCs) [10] and Unified Modeling Language (UML) [11]. For scenario representation we start with scenario stories (modeled with help of templates [12,13]), which we map to interaction diagrams (MSCs).

We detail the TBA framework in the following text. In section 2 we present the SP approximate (used to simplify / prepare the SPs for the techno-business analysis), and explain the main phases of the TBA. In section 3 we discuss some additional analytical opportunities the TBA framework can offer. Namely, model-based mapping and projection techniques can also be used for qualitative and quantitative comparison of various, business, system and technological solutions. Section 4 exemplifies the use of the TBA on a proof-of-the-concept case, while section 5 concludes this work.

2   Techno-Business Modeling Framework

The TBA framework is based on: *(1)* a service platform approximate, *(2)* scenario-based modeling and simulation of SP's technical and business processes and *(3)* portfolio analysis and valuation of services and enablers. Universality of the TBA is based on uniting these analytical techniques in a common framework, which can be used for the techno-business analyses of services and SP solutions, varying from simple service provisions to provisions of complex service and enablers' portfolios. The TBA enlightens simultaneously roles and responsibilities (specified by interactions and collaboration scenarios) of SP entities and business actors: services, enablers, end-users and service and enabler providers. This requires well structured and methodical analyses, which we discuss in the following subsections.

2.1 Service Platform Approximate

A structural service platform approximate, so called Generic Service Platform Model (GSPM) [1,2], hides a complexity of the SPs, simplifies system and technical designs and prepares them for techno-business analyses. The GSPM, shown in Fig.1, abstracts the service platform designs by the following entities (classified according to the set of both technical and business criteria [1,2]): services, enablers, service platform mechanisms (e.g. service discovery, composition, brokering and mediation), service platform capabilities and resources. Due to a high abstraction level the GSPM is also used as a framework for comparison of technical and business solutions, as discussed in section 3. We can conclude that we approximate the SP by two portfolio sets: *(1)*

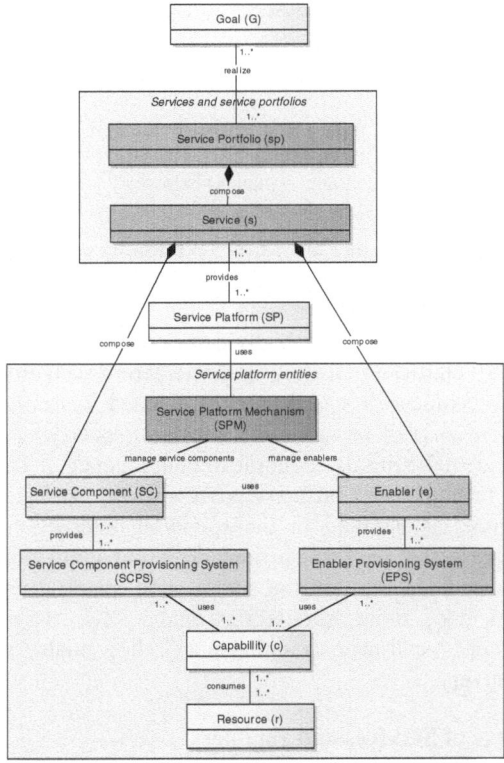

Fig. 1. Generic Service Platform Model (GSPM)

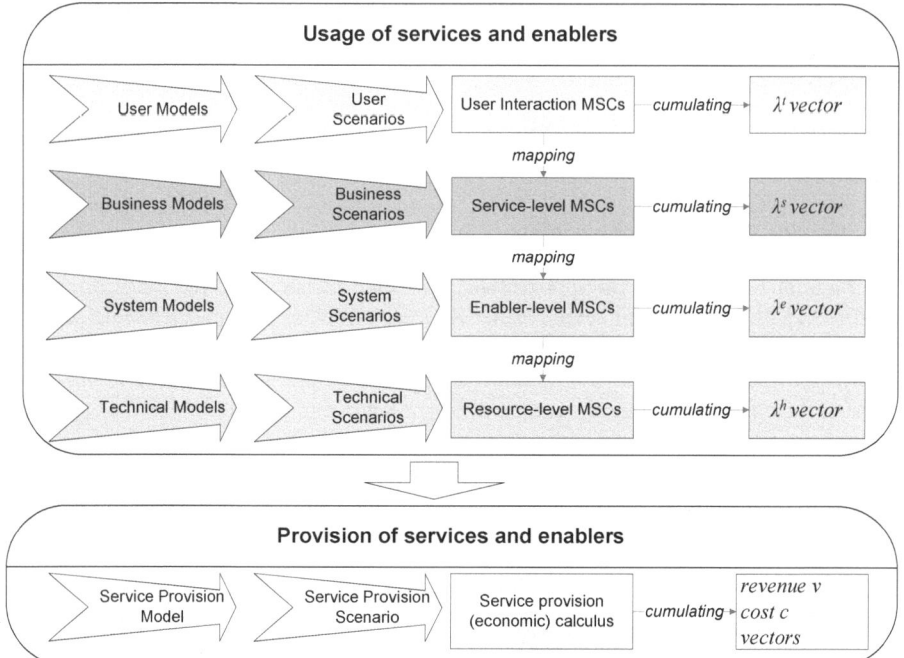

Fig. 2. Techno-business modeling approach (contains two major phases: *usage* of services and enablers, and *provision*). Scenario mapping and projection (cumulating) techniques are illustrated in each phase of the model, and explained below.

end-user service portfolios (in further text service portfolios), and *(2)* portfolios of enablers and service components, used to compose the end-user services and their portfolios.

When the SP is approximated with help of the GSPM (Fig.1), we engage in the TBA. As illustrated in Fig.2, the TBA is divided in two phases: *(1) usage* of services, enablers and other SP entities (analyzing user interaction scenarios, business process, system process and technical process scenarios, needed for service composition and delivery) and *(2) provision* of services and enablers (specifying business aspects of service provision and analyzing its economical consequences). These two models are used to "simulate" technical and business processes and estimate their business and technical performance, as detailed in the following subsections. After these two models have been presented, we discuss scenario-based simulation of periodic use of services and enablers, and focus on the service and enabler portfolio analysis and valuation. To stress once more, simulations analyze two usage aspects: *(1)* how service portfolios support end-user scenarios and *(2)* how enablers are used in service composition and delivery.

2.2 Modeling Usage of Services and Enablers

Scenario Modeling and Mapping
As mentioned before, we engage in process modeling (sketched in Fig.1), starting with user scenarios u_u (indexed by $u = 1 : N_u$), focusing on user activities and

interactions t_m (indexed by $m = 1 : N_t$). We continue with business process scenarios (specifying the support of services s_j (indexed by $j = 1 : N_s$)), system scenarios (determining the usage of enablers e_i (indexed by $i = 1 : N_e$)), and finish with technical scenarios (mapping the system process scenarios to various technological solutions and its resources h_l (indexed by $l = 1 : N_h$)).

Firstly we identify *a set of user / usage scenarios u_u*, representing the typical usage patterns (aiming at realizing user goals). Scenarios, modelled as template-based scenario stories, are mapped to MSC-based usage interaction sets, i.e. sequences of interactions invoked by the user (Eq. 1). Some of these interactions t_m can be supported by the SP functionality (services and enablers). Not all user interactions result in service interactions. Careful scenario analysis has to resolve these mapping issues.

$$u_u \xrightarrow{map} msc\,(t_1,\cdots,t_{N_t}) \xrightarrow{cum} \lambda_u^t = (\lambda_{1u}^t, \lambda_{2u}^t, \cdots, \lambda_{mu}^t, \cdots, \lambda_{N_tu}^t) \tag{1}$$

The λ parameters are cumulative message counts, summing up the number of message instances of the same message type as in a MSC diagram (Eqs. 1-4), convenient for further processing. In such a way we project the MSCs into their message totals, as sketched in Fig.1.

Mapping scenario interactions to services
It should be noted that a user interaction t_m could be mapped to multiple services s_j, representing alternative service supports. E.g. it is possible to send an instant message by various solutions, depending on the service context (SMS, MMS or e-mail). Mapping and cumulating scenario interactions t_m to service invocations s_j is shown in Eq. 2.

$$msc\,(t_1,\cdots,t_{N_t}) \xrightarrow{map} msc(s_1,\cdots,s_{N_s}) \xrightarrow{cum} \lambda_m^s = (\lambda_{1m}^s, \lambda_{2m}^s, \cdots, \lambda_{jm}^s, \cdots, \lambda_{N_sm}^s) \tag{2}$$

Mapping services to enablers
Enablers provide the SP functionality to the services and are measured in units of functionality (invocation counts). The service composition mechanisms [1,2] determine to a great extent the service and enabler level MSCs (Eqs. 2 and 3). In praxis the service composition can be a static software system (producing the same service and enabler composition sets in all scenario instances), or a dynamic (and respond to service context changes or changes in service domains (e.g. availability of enablers and services in service domains. Service composition might vary, resulting in different coefficients in Eq. 3. Each service s_j, is represented by an enabler MSC, representing a set of interactions that enablers e_j and service platform mechanisms must complete in order to deliver service s_j. So we can approximate the *service s_j* as a composition of a set of service components and *enablers e_i* expressed by:

$$s_j \xrightarrow{map} msc\,(e_{1j},\cdots,e_{N_e j}) \xrightarrow{cum} \lambda_j^e = \left(\lambda_{1j}^e, \lambda_{2j}^e, \cdots, \lambda_{ij}^e, \cdots, \lambda_{N_e j}^e\right) \tag{3}$$

s_j is mapped to enablers by using the vector λ_{ij}^e – representing the needed amount of enablers for delivering the service s_j.

Mapping enabler level MSCs to SP capabilities and resources
Each enabler requires a dedication of a set of capabilities, which in turn consume resources, expressed by:

$$e_i \xrightarrow{\;map\;} msc\,(h_{1i}, \cdots, h_{N_h i}) \xrightarrow{\;cum\;} \lambda_i^h = \left(\lambda_{1i}^h,\, \lambda_{2i}^h,\, \cdots, \lambda_{li}^h, \cdots,\, \lambda_{N_h i}^h\right) \quad (4)$$

Such model transformations give us a possibility to specify interaction sets in such a way that they can be analyzed: qualitatively (including structural model analyses), and quantitatively - with help of various projections, e.g. financial (cost and revenue analysis) and resource and capability projections (analysis of usage of service platform capabilities and resources).

Simulation of Periodic Usage of Services and Enablers

Simulation of periodic use of services and other SP entities (enablers and resources) is the next step in our techno-business analysis. Important inputs are vectors specifying: contribution of services λ_{jm}^s to scenarios, enablers λ_{ij}^e to services and consumption of resources λ_{li}^h (by enablers). All the scenario instances are cumulated into the vectors $\lambda_m^s, \lambda_j^e, \lambda_i^h$, representing the totals of their contributions in scenarios. When projections are calculated for single instances of scenarios we continue with their periodisation. Interesting period units for this work are days, weeks, months and years. Simulation of periodic use of services, enablers and resources (denoted as *use* in Eq. 5) is made in the following way.

$$use_{period} = \sum_{subperiod} use_{subperiod}$$

$$use_{subperiod} = \sum_{u} w_u \cdot use_u \quad (5)$$

$$use_u = f(u_u, \lambda_j^s, \lambda_i^e, \lambda_l^h)$$

Where w_u is a weighting coefficient describing usage variation in scenario u, compared to the normative use, while *period* and *subperiod* are period units. We proceed toward valuation of roles, responsibilities and contributions after completion of simulations. A modeling entity under consideration determines the most convenient valuation method. In this analysis we focus on services and enablers. For them a cash flow-based valuation is the most appropriate. We explain it in the following section.

2.3 Modeling Service Provision

Business characteristics of services and service portfolios (e.g. business scale and scope parameters, network effect features, pricing, utility and price elasticity, service life-cycle effects, revenue and cost structure of services) have to be included in the techno-business models and scenarios. Majority is included either in service usage modeling or in service provision modeling (cash flow – based analyses of various business scenarios), which is shortly discussed in this subsection. The commercial service life-cycle is divided in three zones [9]: growth, maturity, and decline. In order to get realistic modeling estimates various information sources are combined, such as: market analyses, field work (empirical studies, field trials), survey techniques and expert analyses. In the backbone of all these analyses is a simple service provisioning model, shortly presented in the text below (described in [1,3]).

A service j generates a revenue v_j per unit of service. It is assumed that revenue v_j generated by a unit of service j is distributed among the actors who participate in creation of service j, performed using a vector of revenue shares γ_j. As detailed in [3] the expected return r_i of the service portfolio (containing M services s_j) is given by Eq.6, where x_{ij} specifies the amount of enabler e_i dedicated to provision of service s_j, and c_i includes both variable and projected/discounted fixed costs of enabler provision. More about the model is given in [1,3].

$$\bar{r}_i = \sum_{j=1}^{M} x_{ij} \left(\gamma_{ij} E\left(\frac{v_j}{c_i \lambda_{ij}}\right) - 1 \right) \tag{6}$$

3   Scenario-Based Analysis and Valuation

Model-based mapping and projection techniques can also be used for qualitative and quantitative comparison of various user, business, system and technological solutions. When *single scenario instances* or classes are analyzed the vectors λ^s_s, λ^e_j and λ^h_i are used. However, often the analysts direct their attention to the scenario sets (scenario trees), representing the complete solutions / designs, as discussed in this section. In these cases they group scenarios according to various modeling and analytical criteria, and develop the *scenario group representatives* λ^s, λ^e and λ^h. Scenario trees, illustrated in Fig.2, are used to compare alternative scenario branches (each with its own set of λ^s, λ^e and λ^h vectors), the difference of which we compare by traversing the branches and applying mappings (*map*) and projection techniques (*project*). In such a way we can compare alternative technical and business designs, as exemplified in the text below.

Comparison of alternative service supports of a user scenario
Various service portfolios can support the same user scenarios. That might change a user experience, utility of services and scenarios, quality of service, and overall value of the service support, but if the difference from the user expectations is not too big, these service portfolios could be considered as alternative service portfolios, capable of supporting the same usage scenarios. Alternative service portfolios can imply alternative business collaborations (resulting in different business roles and responsibilities for business actors). We can assume that the revenue share part for a business actor corresponds to the roles / responsibilities in the service provision, which can be interpreted by the amount of delivered / provided SP entities (services, enablers and resources), in these business and system collaborations. An example comparison of alternative service support of the same user scenarios will traverse two branches with the following nodes in Fig.2: *(a) 1→3→7→12→17* and *(b) 1→4→8→13→18* and analyze them by the following mappings and projections:

$$\lambda^t \xrightarrow{\ map\ } \lambda^s \xrightarrow{\ map\ } \lambda^e \xrightarrow{\ map\ } \lambda^h \xrightarrow{\ project\ } r \tag{7}$$

Comparison of various system collaboration patterns
Another interesting analysis can be a comparison of various system solutions, modeled as various system collaboration patterns (scenarios). Vectors λ^e represent various system solutions for delivering the same service (so the same service can be

composed of different sets of enablers or components). It can in some cases also represent various technical solutions, where some enablers are replaced with others. Alternative service implementations (by alternative enabler portfolios) in Fig. 2 are represented by the following scenario paths: *(a)* $2\rightarrow5\rightarrow9\rightarrow14$ and *(b)* $2\rightarrow6\rightarrow11\rightarrow16$, and analyzed by the following mapping and projections:

$$\lambda^s \xrightarrow{\ map\ } \lambda^e \xrightarrow{\ map\ } \lambda^h \xrightarrow{\ project\ } r \qquad (8)$$

Comparison of various technological realizations
Alternative technical implementations (technological variants) of enablers can also be compared with help of enabler-level MSCs and corresponding cumulatives, as illustrated in Fig. 2 by two branches: *(a)* $5\rightarrow9\rightarrow14$ and *(b)* $5\rightarrow10\rightarrow15$:

$$\lambda^e \xrightarrow{\ map\ } \lambda^h \xrightarrow{\ project\ } r \qquad (9)$$

4   Practical Case – SPICE Mobile Service Platform

We have used our scenario-based modeling approach in several practical cases. Here we will present one of them: a third party service provision platform – based on the SPICE service platform technology [1,7]. Four business actors collaborate in service provision. Each business actor is a separate business unit (enabler or service component provider), as illustrated in Fig. 3: service provider a_1 (responsible for: e_1 – service composition and delivery enabler, and e_3 – A4C enabler), context provider a_2 (providing e_2 – service context enabler), network provider a_3 (delivering e_4 – network enabler) and content provider a_4 (responsible for e_5 – content enabler). The SP model

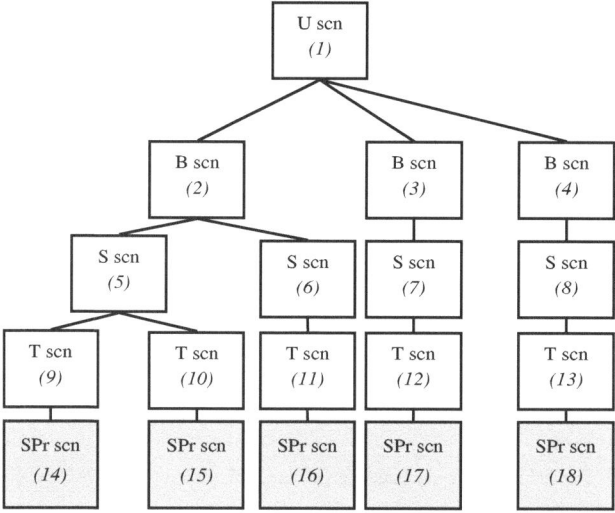

Fig. 3. Scenario tree, containing: user scenarios (*U scn*), business process scenarios (*B scn*), system process scenarios (*S scn*), technical process scenarios (*T scn*), and service provision scenarios (*SPr scn*). The nodes 1-13 focus on usage or services and enabler, while the nodes 14-18 focus on the business aspects of service provision.

contained service portfolios of 6 services, composed of 5 enablers. 5 user groups interacting in 8 usage scenarios have been simulated. Periodic service usage was simulated and calculated for a period of 10 years (anticipated service life cycle). Usage variations on daily, weekly and seasonal basis have been included in the simulations. Business case was limited to Norwegian telecom market and Norwegian users. Business risk estimates were obtained from comparable projects and service offerings. Valuation was based on cash flow projections, where various total and partial net present values have been calculated.

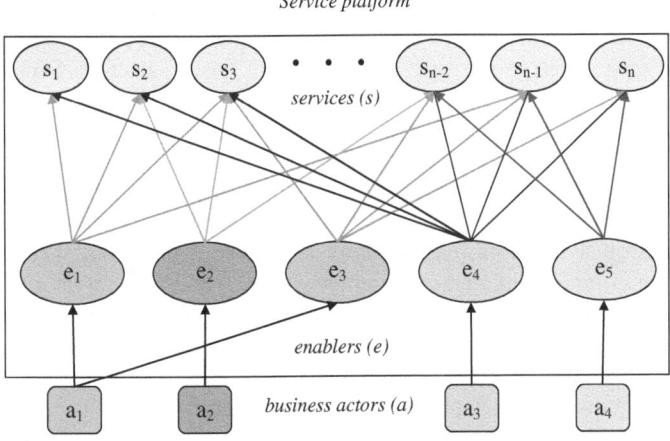

Fig. 4. Sketch of the SP entities, actors and their techno-business management structure

In the analyzed platform one part of enablers is already implemented, while the rest will be either developed (need investment estimation) or rented from other service providers. Three enablers required significant investment: service composition and

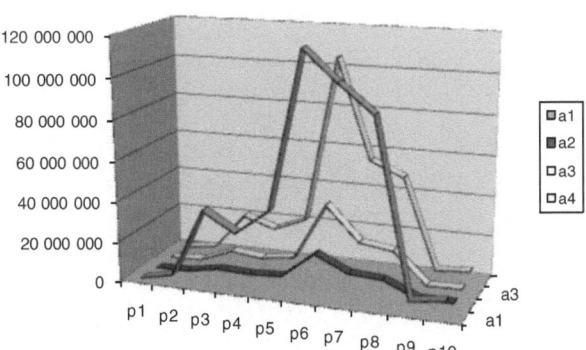

Fig. 5. Income profile for business actors a_i, providing one enabler each (as sketched in Fig. 4), with exception of service provider a_1, who provides two enablers: e_1 – service composition and delivery enabler, and e_3 – A4C enabler. The revenue units are anonymous. Anticipated commercial service life-cycle is 10 years ($p_1 - p_{10}$).

delivery enabler, and improvement of the existing context provisioning system. Investors typically seek the quantitative answers to the following questions. Which services users need in their scenarios (and to which extent)? Which enablers are used in services and to which extent? Are these scenarios equally profitable for enabler providers as for service providers? Which alternative service portfolios might support the same user scenarios? In the text below we give examples of results (with anonymized revenue values).

Income profile for business actors a_i is shown in Fig. 5 (corresponding well to income per enabler they provide). Namely, business actors provide one enabler each (as sketched in Fig. 5), with exception of service provider a_1, who provides two enablers: e_1 – service composition and delivery enabler, and e_3 – A4C enabler. Fig. 5 shows significant difference in cash flow profile of business actors. Service provider (a_1) and content provider (a_4) have much higher cash flow positions. It might indicate the business potential of delivering and integrating service platform solutions (and mechanisms), assuming that our pricing and revenue share mechanisms are used. It is still very unclear how the service discovery, composition, brokering and mediation (service platform mechanisms) will be priced. We believe that they should be treated equally to the other enablers (e.g. network and content provision) because of their importance for composition, delivery and provision of end-user services. At last, they wrap the SP functionality and provide it to the end-users.

Fig. 6 compares the income per user group, based on calculating the value of service support in user scenarios. The differences in profitability of various user groups can be noticed, and it partly reflects the importance and the role of service support for their activities (expressed as interaction patterns). User groups differ in: the services they choose, the way they use them, and the consumed quantities.

Fig. 7 shows differences in the income of various services. When we analyze Figs. 5-7 we have to keep in mind that the choice of service portfolio is not determined just by the business criteria. Additional important moments can play significant role: user preferences, service context, required services / enablers, and even government

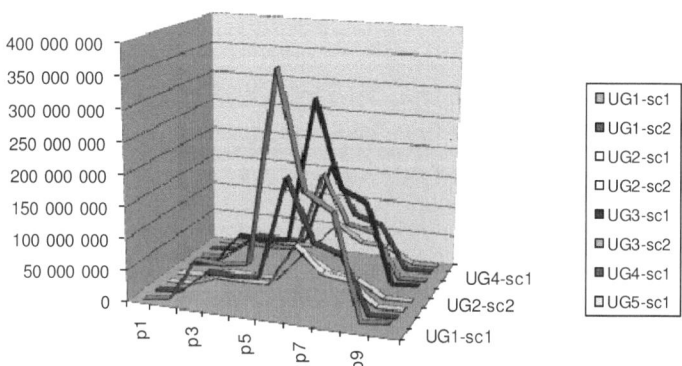

Fig. 6. Contribution of 4 user groups (UG_i) and their 8 usage scenarios (sc_i) to revenue creation (10 years service life-cycle). The revenue units are anonymous. Anticipated commercial service life-cycle is 10 years ($p_1 - p_{10}$).

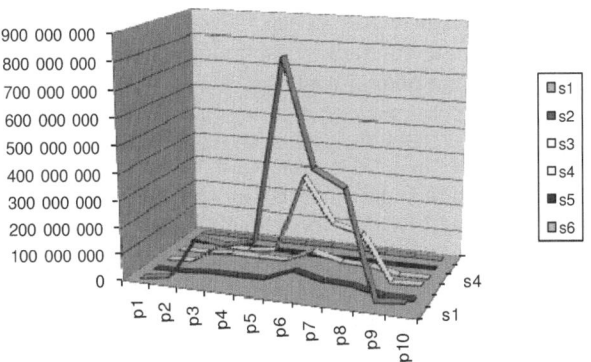

Fig. 7. Contribution of 6 services ($s_1 - s_6$) to revenue creation (10 years service life-cycle). The revenue units and service names are anonymous. Anticipated commercial service life-cycle is 10 years ($p_1 - p_{10}$).

regulation. Very often just a part of the service portfolio elements are profitable, some of them even create just costs, however their total contribution should result in a positive position for the business actor [3].

The TBA approach can also offer the sensitivity information for various entities: services and service portfolios, enablers and enabler portfolios, business actors, user groups and their usage scenarios, various business parameter sets (e.g. pricing schemas, cost factors and revenue models). This information complements well the business model information and gives an insight in some aspects of the risk of business actors.. This result requires that the scenario trees represent well the possible technical and business situations.

As this example and the theoretical part of this paper show, the TBA is a complex analysis, strongly dependent on realistic input coming from four domains (user, business, system and technical). With this limitation/assumption in mind, we believe that it offers a reasonable framework for integrating scenario information in a systematic quantitative analysis. Its further strength and practical relevance lies in the choice of standard and well accepted analytical techniques. However, we stress that we consider the TBA as a complement to other methodologies used in business modeling and analysis. It is particularly convenient for the cases of conceptually new, complex designs, without historical data, where the other methodologies exhibit high degree of speculation, and where some quantitative estimates might help.

5 Conclusion

We believe that the value of business models increases when supported by well founded quantitative (e.g. financial) estimates. This is particularly needed for the conceptually new designs, without historical data, and with difficulties in using comparables. It is difficult to create technical and business analyses for complex, distributed, heterogeneous and pervasive services. The TBA framework might offer some of the answers and estimates. The whole TBA is quite a complex analysis, strongly dependent on realistic input, coming from various analytical domains. Its strength and practical relevance lies in the choice of standard and well accepted analytical techniques. We are continuing improving the approach and simplifying it for the practical use in service and SP analyses.

Acknowledgement

Part of this work is supported and financed by the ISIS project (Infrastructure for Integrated Services), financed by the Norwegian Research Council (NFR # 180122). Author is grateful for the support.

References

1. Zoric, J., Strasunskas, D.: Techno-Business Assessment of Services and Service Platforms: Quantitative, Scenario-Based Analysis. In: Cunningham, P., Cunningham, M. (eds.) ICT-Mobile Summit 2008, Conference Proceedings, IIMC International Information Management Corporation (2008) ISBN: 978-1-905824-08-3
2. Zoric, J.: Practical quantitative approach for estimating business contribution of enablers and service platforms. In: Proceedings of ICIN 2008, Services, enablers and Architectures Supporting Business Models for a New Open World, Bordeaux (2008)
3. Gaivoronski, A.A., Zoric, J.: Evaluation and Design of Business Models for Collaborative Provision of Advanced Mobile Data Services: Portfolio Theory Approach. In: Proceedings of the 9th INFORMS Telecommunications Conference, Telecommunications Modeling, Policy, and Technology. Springer Verlag's Book of Conference Proceedings, Springer, Heidelberg (2008)
4. Gordijn, J.: Value-based Requirements Engineering – Exploring Innovative E-Commerce Ideas. PhD thesis, Vrije Universiteit, Amsterdam (2002)
5. Akkermans, H., Baida, Z., Gordijn, J.: Value Webs: Ontology-Based Bundling of Real-World Services. IEEE Intelligent Systems 19(44), 23–32 (2004)
6. Osterwald, A., Pigneur, Y.: E-Business Model Ontology for Modeling E-Business. In: Proceedings of the 15th Bled Electronic Commerce Conference E-Reality: Constructing the E-Economy, Bled (2002)
7. Ballon, P., Gaivoronski, A., Walravens, N., Zoric, J.: Structural and Quantitative Evaluation of Multi-Actor Business Models for Mobile Service Platforms. In: Proceedings of ICT-Mobile Summit (2008) ISBN: 978-1-905824-08-3
8. Weill, P., Vitale, M.R.: Place to space: Migrating to E-Business Models. Harvard Business School Press, Boston (2001)
9. Kotler, P., Armstrong, G.: Principles of Marketing, 10th International edn. Pearson Education Limited (2004) ISBN 9789582850972
10. ITU-T. Languages for Telecommunication Applications – Message Sequence Chart (MSC), ITU-T Recommendation Z.120, Geneva (1999)
11. Jacobson, I.: The Use Case Construct in Object-Oriented Software Engineering. In: Carroll, J.M. (ed.) Scenario-Based Design: Envisioning Work and Technology in System Development, pp. 309–336. John Wiley and Sons, Chichester (1995)
12. Leite, J.C.S.P., Hadad, G., Doorn, J., Kaplan, G.: A Scenario Construction Process. Requirements Engineering 5, 38–61 (2000)
13. Schneider, G., Winters, J.: Applying use cases: a practical guide. Addison-Wesley, Reading (1998)
14. Pateli, A., Giaglis, G.M.: A framework for understanding and analysing E-Business models. In: Proceedings of the 16th Bled Electronic Commerce Conference - E-Transformation, pp. 329–348 (2003)

The Borders of Mobile Handset Ecosystems: Is Coopetition Inevitable?

Gaël Gueguen and Thierry Isckia

Groupe ESC Toulouse
Télécom & Management SudParis
g.gueguen@esc-toulouse.fr,
thierry.isckia@it-sudparis.eu

Abstract. Today, the mobile phone industry witnesses important changes, shifting from a value chain to a burgeoning business ecosystem. This paper deals with the relationships that are at the very core of mobile OS ecosystems for IMTs (smartphones and PDA): Microsoft-OS, Symbian-OS, Palm-OS and RIM-OS over the period 1998-2006. Our study confirms that an ecosystem's borders are unclear. More than half of our sample' relationships are shared by at least two different ecosystems. The ecosystems we studied do not differ in terms of exclusive relationship which suggests that coopetitive strategies are particularly relevant in mobile platforms war.

Keywords: Ecosystems, Mobile OS, co-opetition, keystone organization.

1 Introduction

When Microsoft launched a new version of Windows Mobile operating system in 2005, Bill Gates spoke in the following terms: "The idea is to create a real ecosystem, with operators, manufacturers and developers[1]". In 2007, Google revealed its broader mobile strategy and released Android, a Java-based operating system that runs on the Linux 2.6 kernel. Android was announced under the Open Handset Alliance, a group of around 30 technology and mobile industry leaders. Under Google's leadership, these companies will work together to create both a more open cellphone environment and a better customer experience, turning cellphones into powerful mobile computers. However, this strategy is far from new. Indeed, ten years earlier Nokia used the same strategy, partnering with major players such as Psion, Motorola, Matshushita-Panasonic, Siemens, Sony-Ericsson and Samsung in order to develop the Symbian operating system. Looking back, Microsoft, Nokia and Google with their respective operating systems for intelligent mobile terminals (IMT), were going down the same road. Their strategies have common features. Basically, they commit resources to get the leadership in the mobile phone landscape, offering a standardised technology (OS) thanks to a wide range of relationships between various players from different sectors, whether they are partners or competitors. Such value webs are an opportunity for different key players to promote their flagship OSs. They are also known as 'business

---

[1] La Tribune, 12 May 2005.

C. Hesselman and C. Giannelli (Eds.): Mobilware 2009 Workshops, LNICST 12, pp. 45–54, 2009.
© ICST Institute for Computer Sciences, Social-Informatics and Telecommunications Engineering 2009

ecosystems" [1]. Teece [2] defines business ecosystems as "a community of organizations, institutions, and individuals that impact the enterprise and the enterprise's customers and suppliers". This paper will focus on the relationships that are at the very core of mobile OS ecosystems for IMTs (smartphones and PDA): Microsoft-OS, Symbian-OS, Palm-OS and RIM-OS over the period 1998-2006. Indeed, these four OSs were the most popular mobile platforms during the study period. In this paper, we analyze the main characteristics of these four business ecosystems. We will focus on "exclusiveness" i.e. the tendency to associate with only a select keystone organization. Given the existence of various players and a large number of potential relationships, the choices pertaining to this exclusiveness and their outcomes are particularly relevant.

1.1 It's All about Business Ecosystems

The key players in the ICT field (Apple, SAP, Cisco, IBM, Symbian, Microsoft, etc.) often use the concept of "business ecosystems" to define the loose networks of suppliers, distributors, outsourcing firms, complementors, technology providers, that affect, and are affected by, the creation and delivery of a company's own offerings. For a specific company, it is very important to join such a network in order to benefit from business opportunities [3]. The ecosystem-based view is a very exciting framework which provides an alternative interpretive lens for better understanding new forms of dynamic networked co-operative business processes [1], [4]. In his seminal book Moore [1] sees business ecosystem as "a community of businesses and individuals that co-evolve, sharing one or more resources on the basis of a common strategic destiny". This concept relies on different theoretical approaches [5]. Shapiro and Varian [6] assert that because of the compatibility between certain technologies, businesses that sell complementary products or services have to develop relationships with their allies. Thus, forming alliances, cultivating partners, and ensuring compatibility (or lack of compatibility) are critical business decisions especially in the ICT sector where standards are an important issue.

Relationships between the firms of an ecosystem are complex and show a mix of cooperation and competition, illustrating situations of coopetition as analysed by Nalebuff & Branderburger [7]. Because of this, the frontiers of an ecosystem are unstable and keep changing depending on the interactions between member firms. An ecosystem is a business community which brings together firms from various industries which are interdependent. These business communities are usually structured around a leader, which strives to share its commercial philosophy or its technological standard [8]. Moore [9], [10] also stresses this dimension and the need for the leader to develop the kind of vision to which the ecosystem's members can adhere. In this framework, the role of the leader is to encourage the convergence of all the other community members' vision and ensure that their efforts will enable the development of beneficial synergies for the customers. This shared vision is indeed a way of structuring innovation and ensuring coordination amongst actors within the ecosystem. The ICT sector is closely related to the concept of the business ecosystem because it is made up of very dynamic interdependent markets [11]. Indeed, in such dynamic markets, leaders or "would-be" leaders often try to introduce standards that will ensure market stability and their market dominance. But for such a stability to emerge, a standard must be introduced and widely accepted. As a consequence, we can observe a mix of competition and collaboration between companies which leads to great instability in the early phases of the ecosystem life-cycle. In the mobile phone sector, navigating business ecosystems is very important for those companies wishing to promote a standard mobile OS and achieve sustainable growth. For instance, a common strategy for mobile platform providers is to build a software marketplace, encouraging a large developer community that will increase the penetration rate of their OS. The success of a developer program not only depends on the software marketplace but also on the platform's health i.e. the OS ecosystem. Third-party developers are interested in making money and creating great applications. In the first place, they will choose the platform that lets them easily create these applications. However, if they can't make money from a platform, they will move away to another. At this stage, given the existence of cross network externalities, an important issue is how big is the user base?

1.2 Is Exclusivity in Relationships Possible?

In previous research, we suggested that coopetitive relationships have a great influence on smartphone and PDA ecosystems [12]. However, our research focussed on the direct relationships between focal firms or keystone organizations and did not lead to a quantitative analysis. In order to better understand the main differences between rival ecosystems, we are now going to focus on the relationships we identified in such ecosystems. For instance: does the number of relationships within a business ecosystem increase over time? Are these relationships based on the same incentives (commercial agreements, long-term partnerships etc.)? Do these ecosystems differentiate themselves through their members' activity? At this stage, we are mainly interested in the specificity of the relationships' evolution, the very nature of these relationships and the type of actors that are involved in these relationships. At the end, we hope the answers obtained could help us better understand how context-specific the exclusivity of such relationships is. Basically, we want to know if a specific business ecosystem can be made up of relationships between players that are not in touch with the keystone of a rival ecosystem. In other words, what is the degree of coopetition [7] between rival ecosystems? Coopetition refers to the collaborative arrangements of two or more competitors while at the same time these firms compete at the market [13]. Hence, coopetition builds on the idea that competitors should not just be considered as rivals for market dominance but also as valuable sources of innovation. We have to appreciate coopetition both within and between ecosystems. For instance, Figure 1 below describes two different ecosystems (A and B) consisting of a group of relationships between firms X, Y and Z (1). Firms X are the "leaders" [1] or the "keystone organizations" [4] in their respective ecosystems. These firms are in competition (2a). The ecosystems A and B are also in competition in order to promote their mobile OSs as a dominant design (2b). However, keystone organizations X can build direct relationships (3), which refer to coopetition strategies. When different players belong simultaneously to both ecosystems (Z) there will be an indirect coopetition.

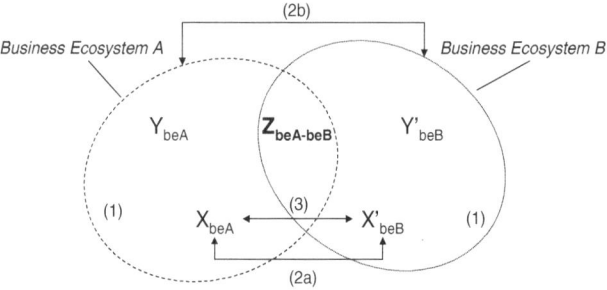

Fig. 1. Business ecosystems and keystone organizations

An examination of the academic literature in the field of business ecosystems reveals that the question of "exclusiveness" is still underdeveloped [1], [14], [4], [2]. Since firms can contribute in different ecosystems, these contributions can reduce resources specificity between ecosystems. In respect to this question, the number of non-exclusive relationships within business ecosystems is particularly relevant. In the same vein, are there differences in the appeal to forming relationships with exclusive actors?

2 Methodological Approach

First of all, we postulate that:

- Two firms that initiate a partnership belong to the same ecosystem; their respective interests in this partnership will converge meaning that their destinies are partially linked. Thus, a firm in connection with a player that works on a specific mobile platform belongs to this mobile OS ecosystem. Of course, it's about a wide approach of ecosystems and some of these relationships may be more important than others.

- Based on secondary data we extracted from professional journals, we listed various relationships between players in the mobile landscape in order to build an exhaustive sample. This sample can be used in order to describe (characteristics) the spectrum of relationships making inferences from sample data to the population. We have chosen French professional journals to facilitate access to raw data.

We reviewed three main French journals (Les Echos, La Tribune and 01net.) over the period 1998-2006 in order to extract each article dealing with mobile OS platforms we were interesting in. We only selected articles that included the names of the key players associated with these platforms. We collected about a thousand articles. Once filtered, we retained 738 articles that were closely related to our topic. At the end, we identified 237 collaborative relationships between 96 companies (OEM, ODM, ISV, Content providers, MNO etc.) and 4 keystone organizations. Then, we built an adjacence matrix in order to map possible cooperative links with key players. These links were selected according to their importance. After this stage, we used social network software (Ucinet / NetDraw) in order to draw from the adjacence matrix different sociograms depending on the nature of the relationships. Analysing the multiple cooperative [15] and coopetitive [16] relationships with social network tools has become increasingly commonplace in research. Such a methodological approach makes it possible to analyse complex relationships and to appreciate "degree centrality" i.e. the degree of proximity between key players. This approach allowed us to evaluate cooperation both inside and outside the business ecosystems we focused on. In this context, we were only interested in the links that relate players with keystones organizations in charge of promoting their respective platforms: Microsoft, Symbian, Palm and RIM. Of course, various OSs were available when we conducted our study. However, in order to simplify reality we focussed on the most popular OSs and the major players. The relationships we identified during our study can be sorted as follows:

- *Agreements*: it's a one-off relationship in which players are weakly implicated (EADS and RIM joined forces to deliver security certification of the BlackBerry for governmental organizations in Europe),
- *Customer-supplier relationships*: here it's mainly about licensing (for instance, Sony and other OEMs joined the PALM OS licensee family),
- *Partnerships /Alliances*: these long-term agreements are designed to manage cooperative efforts in creating or exploiting technology (for instance Nokia and Symbian Ltd before the creation of the Symbian Foundation).

Among the 237 relationships we identified (dyadic relationships), there can be redundant relationships especially if a new specific agreement was noticed over the period. In respect to "exclusiveness", there won't be multiple counting. We tried to know if there was at least one relationship between a firm and a keystone organization during the study period. Thus, we identified 160 relationships -among 92 companies- with one of the four keystone organizations.

3 Findings

First of all, we analyzed the number of constitutive relationships by ecosystems in order to appreciate how these relationships evolve year after year. As suggested below (Fig 2), there are variations in the number of relationships. We observed an increase then a decrease of the number of relationships. Whatever the ecosystem, there was no progressive increase in the number of relationships as is the case in a life-cycle curve (Bass Curve). A khi² adjustment test for the relative part of the relationships for each year is not significant (khi² = 1.60 ; DDL : 24).

Then, we tried to analyze if membership of a mobile ecosystem generated specific kinds of relationships. For each ecosystem, we identified a particular form of relationship (Table 1), either based on customer-supplier relationships, alliances or basic partnerships. We run a khi² test that reveals a significant difference (khi² = 32.65 ; DDL : 6 ; p < 0.001). For instance, the Windows Mobile ecosystem mainly relies on customer-supplier relationships based on licensing agreements. The RIM OS ecosystem is also based on the same type of relationships. In the case of Symbian, the ecosystem is based on numerous short-term basic agreements and alliances. Here, it's more about long-term. The Palm-OS ecosystem is similar to Symbian OS.

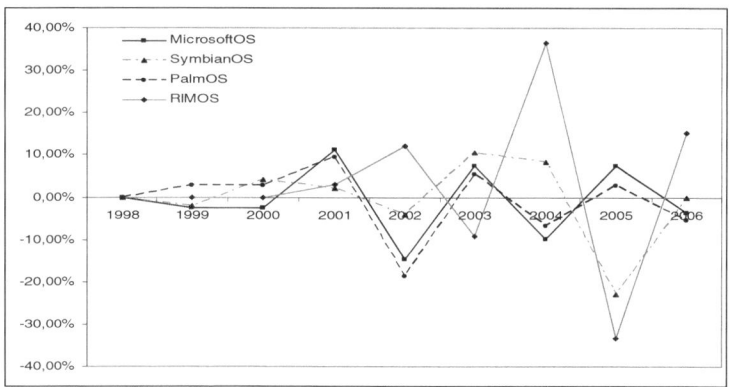

Fig. 2. Ecosystems' evolution over the study period

Table 1. Three types of relationships

Type of Relationship (237)	Microsoft	Symbian	Palm	RIM	Total
Simple Agreement	30.86%	41,67%	41.33%	48.48%	38.82%
Customer-Supplier	55.56%	25%	21.33%	45.45%	37.13%
Alliances	13.58%	33.33%	37.33%	6.06%	24.05%
Total	1	1	1	1	1

Now, we are interested in the species within these ecosystems, *i.e.* members' 'core business' in order to appreciate functional diversity. Indeed, functional diversity affects ecosystem properties and consequently, its sustainability. As suggested in Table 2, we identified 7 different types of players which are quite different from each other (khi² = 52.78 ; DDL = 18 ; p < 0.001). For instance, players within the Windows Mobile ecosystem are mostly mobile handset OEMs and electronic manufacturers and suppliers. However, even if the part of the relationships with these players is the largest within Windows Mobile ecosystem, it is, compared with the other ecosystems, one of the weakest (quite similar to Palm-OS). On the contrary, the part of the relationships with mobile handset OEMs within the Symbian ecosystem is the biggest one (41.67%). This point is consistent with Symbian's history since the company's founders are mainly mobile OEMs. The distribution in the case of Symbian and Palm are quite similar except for the section "Others". Indeed, with Palm we identified a set of relationships with "niche players" that focus on very specific fields in the mobile landscape. In the RIM-OS, the business ecosystem relies on relationships with mobile OEMs and MNOs.

Table 2. Functional diversity

Activities (237)	Microsoft	Symbian	Palm	RIM	Total
OS	6.17%	12.50%	13.33%	12.12%	10.55%
Mobile OEMs	23.46%	41.67%	24.00%	27.27%	27.85%
PC OEMs	18.52%	8.33%	4.00%	3.03%	9.70%
Electronic OEMs & Suppliers	18.52%	8.33%	6.67%	3.03%	10.55%
MNO	18.52%	10.42%	8.00%	24.24%	14.35%
Software / Internet Services	11.11%	14.58%	16.00%	12.12%	13.50%
Others	3.70%	4.17%	28.00%	18.18%	13.50%
Total	1	1	1	1	1

Table 3 indicates for each ecosystem if the relationships take place only within a specific ecosystem or if these relationships are shared, at least with one rival ecosystem. Here, a khi² test indicates that there are no significant differences for a fixed significance level of 0.05 (khi² = 6.60 ; DDL = 3 ; p < 0.09). Thus, it seems that all the ecosystems studied have the same share of non-exclusive relationships. Symbian OS share about three quarter of its relationships with at least one other rival ecosystem. RIM-OS follows the same trend. For both Microsoft and Palm OSs, relationships are divided up in a more equal way between unique and shared relationships. At the end, about 60% of the relationships identified for the four ecosystems are shared relationships. Table 3 is also interesting since it makes it possible to identify the number of rival ecosystems concerned with these sharing relationships. For instance, RIM-OS shares 27.27% of its whole relationships with the three other rival ecosystems.

Table 3. Shared or common relationships

				Shared or Common relations with at least 1 rival OS			
Ecosystem (160)	Unique Relationship	Shared relation with at least 1 rival OS	Total	Shared relation with 1 rival OS	Shared relation with 2 rival OSs	Shared relation with 3 rival OSs	Total
Microsoft	49.02%	50.98%	1	15.69%	23.53%	11.76%	50.98%
Symbian	23.33%	76.67%	1	20.00%	36.67%	20.00%	76.67%
Palm	46.67%	53.33%	1	20.00%	20.00%	13.33%	53.33%
RIM	31.82%	68.18%	1	22.73%	18.18%	27.27%	68.18%
Total	40.54%	59.46%	1	18.92%	24.32%	16.22%	59.46%

Basically, it seems that shared relationships between rival ecosystems are a rule or a common feature. From this point of view, it seems interesting to visualize the relationships between rival ecosystems (Fig 3). For instance, Table 4 reveals that 38.89% of Microsoft ecosystem members are engaged in relationships with the Symbian ecosystem. Figure 3 summarizes all the dyadic relationships between rival ecosystems. In such a context, the Microsoft case is interesting, since the Windows Mobile ecosystem is connected with its rivals in the same proportion, turning the

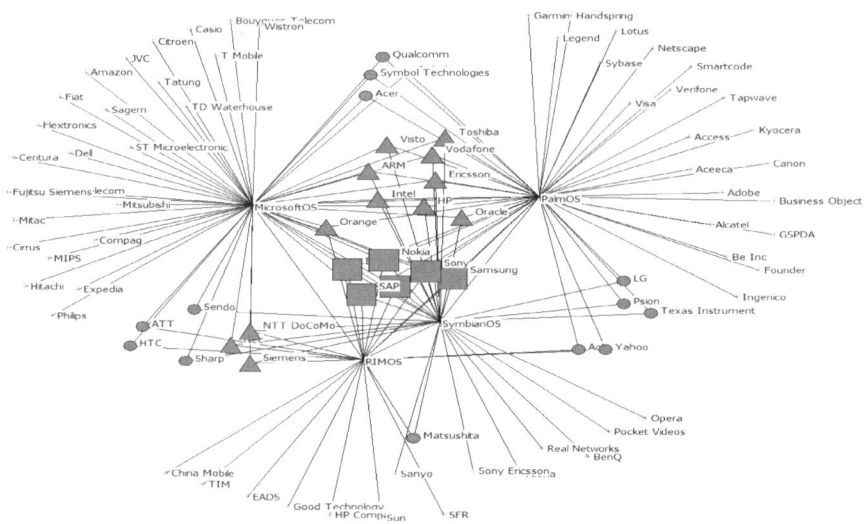

Fig. 3. Ecosystems' map

Table 4. Relationships between rival ecosystems

Ecosystem (160)	Microsoft	Symbian	Palm	RIM
Microsoft	-	38.89%	38.89%	25.93%
Symbian	63.64%	-	57.58%	36.36%
Palm	43.75%	39.58%	-	22.92%
RIM	56.00%	48.00%	44.00%	-

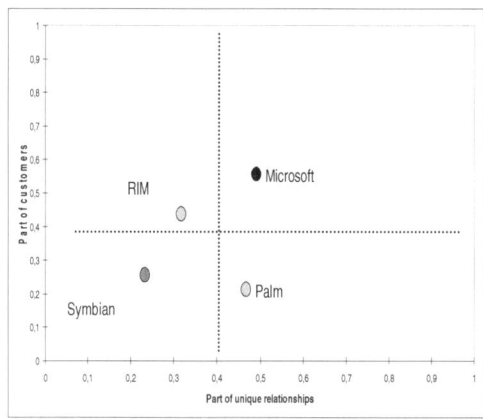

Fig. 4. Part of customers vs. part of unique relationships

platform into a real hub. Basically, figures in Table 4 indicate that the frontiers of an ecosystem are quite unclear.

In Figure (3) below, we mapped relationships between business ecosystems. One can see players that are in touch with the four ecosystems (square), with three ecosystems (triangle), with two ecosystems (circle) and those which are with only one mobile OS ecosystem.

To go further, Figure 4 presents the four ecosystems studied given two quantitative parameters: the part of customers (as opposed to the part of agreements and alliances) and the part of unique relationships (as opposed to shared or common relationships with at least one OS). Data are related to relationships within a specific ecosystem. The matrix below illustrates the importance of commercial relationships (the part of customers *i.e.* OS licensees) for a specific ecosystem (Y) and the specificity of its memberships (X). We notice that the Windows Mobile ecosystem is made of more commercial connections and unique players than the Symbian OS.

4 Conclusion

In this study, we were interested in mobile OS ecosystems. Our study confirms that an ecosystem's borders are unclear. More than half of our sample' relationships are shared by at least two different ecosystems. For instance, 76.67% of Symbian ecosystem members are members of a rival ecosystem. The ecosystems we studied do not differ in terms of exclusive relationship which suggests that coopetitive strategies are particularly relevant in the ecosystem-based view. Beyond the analysis of exclusive relationships between keystone organizations, we suggest that indirect coopetition also characterizes business ecosystems. However, there are several limits to our study:

- We considered that the relationships between players had the same value;
- We did not study relationships between players that were not in the central position (keystones),
- We did not analyze the other ecosystems present over the study period, such as Linux (LiMo);
- We did not identify relationships with regard to their year of appearance to determine if the membership of two ecosystems was simultaneous or not.

In order to go further in the analysis of business ecosystems in the mobile landscape, longitudinal studies should be carried out in order to better appreciate the relationships' evolution across time. Nevertheless, our results confirm the need to develop external relationships between business ecosystems. From this point of view, it seems clear that both the Android ecosystem and the Apple iPhone ecosystem will have to open their doors to external players, whether they are already members of rival ecosystems or not. At the end, the emergence of new forms of hybrid competition that includes competition and co-operation drives the need for relational strategies. The ability to create and manage relationship with a network of collaborators will be a key success factor in the mobile ecosystem war.

References

1. Moore, J.F.: The Death of Competition: Leadership and Strategy in the Age of Business Ecosystems, Harper Business (1996)
2. Teece, D.: Explicating Dynamic Capabilities: the Nature and Microfoundations of Enterprise Performance. Strategic Management Journal 28(13), 1319–1350 (2007)
3. Gulati, R., Nohria, N., Zaheer, A.: Strategic networks. Strategic Management Journal 21(3), 203–215 (2000)
4. Iansiti, M., Levien, R.: The Keystone Advantage: What the New Dynamics of Business Ecosystems Mean for Strategy, Innovation, and Sustainability. Harvard Business School Press (2004)
5. Peltoniemi, M., Vuori, E.: Business ecosystem as the new approach to complex adaptive business environments. In: Proceedings of E-Business Research Forum, Tampere, pp. 267–281 (2004)
6. Shapiro, C., Varian, H.: Information Rules: A Strategic Guide to the Network Economy. Harvard Business School Press, Boston (1998)
7. Nalebuff, B., Brandenburger, A.: La Co-opétition, une révolution dans la manière de jouer concurrence et coopération, Village Mondial (1996)
8. Pellegrin-Boucher., E., Gueguen, G.: How to manage co-operative and coopetitive strategies within IT business ecosystems, the case of SAP, the ERP leader. In: EIASM Workshop on coopetition strategy: toward a new kind of interfirm dynamics?, University of Catania, Italy (September 16-17, 2004)
9. Moore, J.F.: The Rise of a New Corporate Form. Washington Quarterly 21(1), 167–181 (1998)
10. Moore, J.F.: Business ecosystems and the view from the Firm. The Antirust Bulletin, 31–75 (2006)
11. Eisenhardt, K.M., Brown, S.L.: Patching. Restitching Business Portfolios in Dynamic Markets. Harvard Business Review, 72–82 (May-June 1999)
12. Gueguen, G.: Coopetition and business ecosystems in the information technology sector: the example of Intelligent Mobile Terminal. International Journal of Entrepreneurship and Small Business (forthcoming) (2009)
13. Luo, Y.: Coopetition in International Business. Copenhagen Business School Press (2004)
14. Stanley, G.: Management and Complex Adaptation: A research note. Management International 3(2), 69–79 (1999)
15. Dittrich, K., Duysters, G.: Networking as a Means to Strategy Change: The Case of Open Innovation in Mobile Telephony. J. Prod. Innov. Manag. 24, 510–521 (2007)
16. Chien, T.-H., Peng, T.-J.: Competition and Cooperation Intensity in a Network - A Case Study in Taiwan Simulator Industry. Journal of American Academy of Business 7(2), 150–156 (2005)

Trends in Mobile Application Development

Adrian Holzer[1] and Jan Ondrus[2]

[1] University of Lausanne, 1015 Dorigny, Switzerland
adrian.holzer@unil.ch
[2] ESSEC Business School, 95021 Cergy, France
ondrus@essec.fr

Abstract. Major software companies, such as Apple and Google, are disturbing the relatively safe and established actors of the mobile application business. These newcomers have caused significant structural changes by imposing and enforcing their own rules for the future of mobile application development. The implications of these changes do not only concern the mobile network operators and mobile phone manufacturers. This changed environment also brings additional opportunities and constraints for current mobile application developers. Therefore, developers need to assess what their options are and how they can take advantages of these current trends. In this paper, we take a developer's perspective in order to explore how the structural changes will influence the mobile application development markets. Moreover, we discuss what aspects developers need to take into account in order to position themselves within the current trends.

1 Introduction

Mobile computing has caught the attention of the research community for quite some time and has also reached the commercial industry and mainstream consumers via smartphones and PDAs. More than ever, such devices can run rich stand-alone applications as well as distributed client-server applications that access information via a web gateway. This opens new avenues for future mobile application and service development. During many years, the development of mobile services was mostly controlled and managed by the mobile network operators (MNO), phone manufacturers, and some mobile application and content providers. Recently, this has changed with the arrival of new mobile phones and platforms such as the iPhone. Development of mobile applications has generated more interest among the independent and freelance developers. The constant improvement of hardware related to mobile computing (e.g., better computing power, larger wireless network bandwidth) clearly enhance capabilities of mobile devices. The potential of the mobile application market is seen to reach $9 billion by 2011, according to Compass Intelligence[1]. Traditionally, in the mobile application industry, there are several actors intervening along the value chain [1,2,3,4,6,10,15],

---

[1] http://www.compass-intelligence.com/content.aspx?title=PressRelease04

C. Hesselman and C. Giannelli (Eds.): Mobilware 2009 Workshops, LNICST 12, pp. 55–64, 2009.
© ICST Institute for Computer Sciences, Social-Informatics and Telecommunications Engineering 2009

in which each actor has its own importance. The current trends indicate that the market structure and value chain are evolving. Roles are changed, combined and exchanged. Some lost some control on the device (i.e., MNO), some got new revenues streams (i.e., Portal provider), and some became more seamlessly integrated into the platforms (e.g., financial institutions, content providers).

In this paper, we define *platform providers* as providers of operating systems and development tools to enable the creations of high level applications. The current mobile development market is dominated by five big Platform providers namely: Nokia with its Symbian OS (46.6%),[2] Apple with its iPhone OS (17.3%), RIM with its Blackberry OS (15.2%), Microsoft with its Windows CE OS family (13.6%), and LiMo Foundation with its Linux Mobile operating system (5.1%). Furthermore, Google recently launched its Android operating system and is expected to rapidly become part of the big players in the industry. In this paper, we provide a thorough analysis of the current mobile development landscape with hints of future trends as well as indications for developers on what aspects can be used to position themselves in the market. In order to structure our analysis, we propose to describe the current practices by examining the development tools, the different types of portals, and the different levels of platform integrations.

This paper is organized as follows: Section 2 provides a detailed analysis of the current mobile development platform landscape. Then Section 3 points out current trends in the industry. Section 4 analyses the choices independent developers face when deciding for which platform(s) they want to develop. Finally, Section 5 concludes and provides an outlook on future research opportunities.

2   Current Practices

To structure the description of the current practices, we propose to examine the current mobile development platforms from the point of view of individual mobile application developers. We start by classifying the platforms in different categories depending on the three main components depicted in Figure 1. First, the developer uses *development tools* to build its mobile application. Second, the developer publishes its application on a *portal*, from which the consumer can download the application onto its *mobile device*. This model, adapted from [1], includes developers, the application portal, consumers, and all the processes related to the publishing and purchasing of a mobile application.

This model (Figure 1) supports us to separate and examine three main issues, which are addressed in different subsections. In 2.1, we look at the different kinds of *development tools* that are supported. This helps to characterize the type of technology each platform provides for developers (e.g., software development kit). More precisely, we determine if the technology provided has an open access or not (i.e., opensource versus proprietary sources). In 2.2, we describe the different types of *portals* for each platforms. We characterize portals that act as intermediaries between developers and consumers. We differentiate between

---

[2] Percentages represent market shares of the worldwide smart phone shipments in Q3 2008. Source: http://www.canalys.com/pr/2008/r2008112.htm

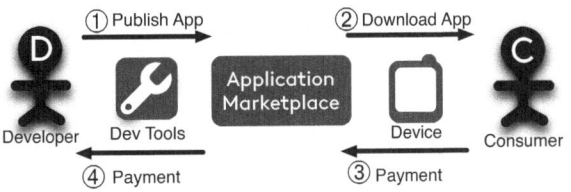

Fig. 1. Mobile application distribution model

centralized and decentralized portals. In 2.3., we look at the level of *integration* of each platform, from no integration to a full distribution model integration.

2.1 Development Tools

Central to every development platform is its software development kit (SDK), which enables third party developers to deliver applications running on the platform. Such a kit can include among other things, libraries, debuggers, and handset emulators. Existing platforms have taken different approaches when it comes to sharing their SDK with developers. Some have chosen to restrict access as much as possible, whereas others have chosen to disclose the entire source code of their SDK and OS. Using the terminology introduced by Raymond in [12], we call *bazaar* an open source platform, where any third party developer can access the entire platform with no, or little, restrictions and contribute to the construction of its structure. Conversely, we call *cathedral* a closed platform, where most of the code is hidden from developers and an all mighty architect plans every inch of the platform's construction.

The cathedral model. Half of the actors, representing roughly half of the customers, chose the proprietary path (Apple, Microsoft and RIM). The other half chose to engage into an open source technology (Linux, Google and Nokia). Proprietary platforms all keep the source code of their SDK and OS hidden from any outsider. The difference between Apple, Microsoft and RIM, is their level of control over what developers can install on the platform. Apple has an almost unlimited control over third party applications since all applications must be approved before they can be released. RIM and Microsoft, on the other hand, are more lenient. Advantages of closed technology for the platform provider include being to sell and control your platform.

The bazaar model. In contrast, open platforms grant developers access to all or parts of the source code of their SDKs and their OS. Among the three open source platforms, Linux seems to offer the most freedom, followed by Google who, for example, denies access to Bluetooth and Instant Messaging APIs for security reasons in their current SDK release. Nokia is in a transition phase making its Symbian OS open source. It is still not clear how open it will become. Benefits of open technology for the platform provider include being able to reduce development and maintenance costs of the platform by taking advantage of the pool of open source developers. Reduced development costs can lead up to reduced platform price and therefore possibly increased number of consumers [13].

2.2   Portals

In order for applications to pass from developers to consumers, an application portal must be created. Mobile portals are considered to be an essential element in the mobile application distribution process since they play the role of intermediary between developers and consumers. Some scholars predicted that the number of portals would increase [4], whereas others predicted that the portal market would consolidate given time [3]. In the current market, both phenomena are present. Some platforms use a *centralized* single point of sale strategy and others use a *decentralized* multiple points of sale strategy.

Decentralized portal. Nokia, Linux, Microsoft, and LiMo have a decentralized portal approach. Developers can freely upload their applications on any third-party portal, as there is no centralized policy. In this model, all portal providers can freely compete in order to gain customers and applications. The downside for the consumer is that the great variety of portals does not provide a comprehensive oversight of existing applications.

Centralized portal. In this model, one portal is proposed as the main portal on which most applications are published. This approach gives the main portal provider a competitive advantage over others. Apple and Google propose such a single point of sale with the AppStore and the Android Market. However, these two platforms have a different approach. Apple pushes for a unique and exclusive portal with a strict application review process. This restrictive approach has led to the creation of alternative "black" portals such as Installer and Cydia. Google, on the other hand, does not restrict the publication of applications to its portal and does not plan to review applications prior to publication.

2.3   Platform Integration

Some platforms focus on their core business, which is to provide an OS with programming support for developers, whereas others integrate the entire distribution process. Hereafter, we classify platforms according to their level of integration similarly to [7], but instead of taking into account the whole value chain, we focus on the distribution process, where we identified four different types of integrations, namely *full integration, portal integration* as well as *device integration* and *no integration* (see Figure 2).

Full integration. Platforms with a full integration have a strict control over every step of the distribution model from device manufacturing to application publishing, as depicted in Figure 2.①. Apple and Nokia exhibit such a strong integration. Apple produces the device on which its OS runs, namely its iPhone, and it owns the unique authorized portal for mobile applications, namely the AppStore. Furthermore, Apple also plays the role of content provider with the

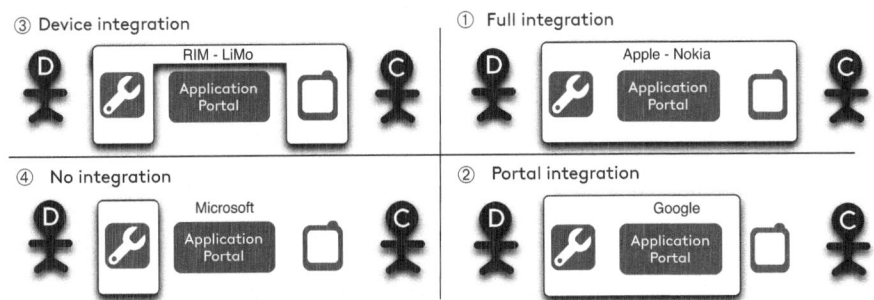

Fig. 2. Platform integration

iTunes store available on the iPhone. Similarly, Nokia manufactures its phones and provides an application portal as well as other content via its OVI[3] portal.

Portal integration. Platforms with portal integration focus on application development and application sale by integrating a portal, as depicted in Figure 2.②. Google provide such an integration with its Android Market. Conversely to Apple, Google does not manufacture the mobile phones on which its OS runs. Nevertheless, they have a strong relationship with a manufacturer on which the platform runs for the moment, namely HTC.

Device integration. In the device integration model, platforms also manufacture devices but are not in the the the application portal business, as shown in Figure 2.③. RIM and LiMo are such platforms. RIM manufactures its Blackberry mobile devices but does not provide a portal. The LiMo foundation can also be considered to follow such a model since it is composed of handset manufacturer such as Motorola, NEC, Panasonic and Samsung.

No integration. Platforms with no integration focus only on their core business as depicted in Figure 2.④. For example, Microsoft does not manufacture mobile devices, nor provide an application portal.

3   Trends

Over the past few years, we have observed that the relatively stable market has evolved in three distinct directions. First, there seems to be a strong trend towards portal centralization. Second, there is an increased number of actors providing open source technology. Third, platforms are moving towards a higher level of integration.

3.1   Towards Portal Centralization

Prior to the introduction of Apple's AppStore and more recently Google's Android Market, platforms did not have a central portal. With the introduction of

---

[3] More information about ovi can be found on: http://www.ovi.com

its AppStore, Apple has proven that a mobile application market should not be underestimated and can represent an important revenue stream. According to CEO Steve Jobs, the AppStore has generated a revenue of a million dollars a day in its first month of existence.[4] There are currently 15000 applications on the portal, which have been downloaded a total of 500 million times. Note that these figures grew by 50% in the last month.[5] Following Apple's lead, traditional platforms like Nokia, RIM and Microsoft seem to be moving in this direction. Nokia is pushing its OVI portal and RIM has developed its own Application Center. Microsoft is also planning to launch its own version of the AppStore called SkyMarket with the next version of Windows Mobile (WM7). Figure 3 depicts this trend.

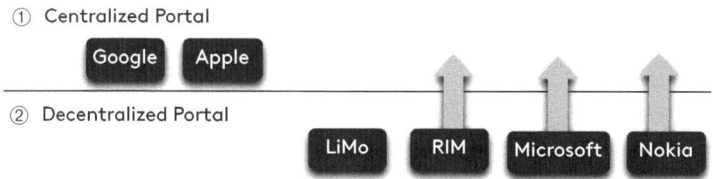

Fig. 3. Portal trends

3.2 Towards Technological Openness

Among the major mobile platforms, LiMo used to be the only player in the open source field. Nokia has moved in this direction after acquiring Symbian OS. Google has also followed this trend. The transition phase from a closed to an open architecture will be critical for the future success of the platform [5]. The shift, depicted in Figure 4, of this major player towards openness means that from a situation with mostly closed systems, we have moved to a situation with a small majority of devices running an open source system. Nevertheless, this shift does not indicate that other platforms will follow. Among the closed platforms, RIM is probably the only one that might go open source, since Microsoft and Apple are strong advocates of proprietary software. So far, it is still hard to evaluate what impact open-source software might have on the current mobile application developments. The successful model that Apple established does not suffer from the proprietary software clauses. The other platforms hope that the open-source option could help them to better compete in the platform war.

3.3 Towards Full Integration

Another trend is the emergence of more integrated platforms, as shown in Figure 5. Before the introduction of Apple's platform, there was no *fully integrated*

---

[4] Wall Street Journal, August 11 2008:
   http://online.wsj.com/article/SB121842341491928977.html?mod=2_1571_topbox.
[5] Businessweek, January 16 2009: http://www.businessweek.com/technology/ Byte-OfTheApple/blog/archives/2009/01/the_app_store_s.html

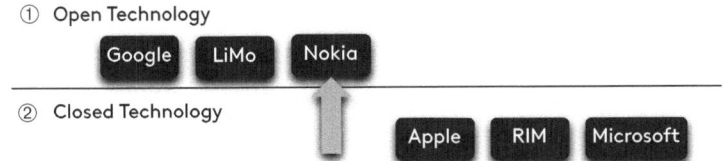

Fig. 4. Technological trends

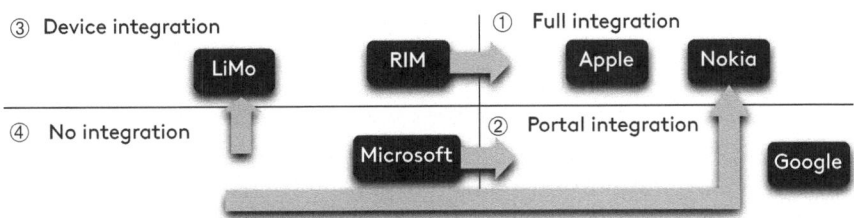

Fig. 5. Integration trends

mobile platform. Moreover, there was no platform with *portal integration* before the introduction of Google's platform. Symbian OS is an example of the trend towards integration since it started as a platform with no integration, before it was integrated by Nokia to become a *device integrated* platform and finally by launching OVI, it became *fully integrated*. RIM is also expected to soon become *fully integrated* with the introduction of its Application Center. Furthermore, with Microsoft moving towards *portal integration* there will be no major platform left without integration. Some scholars have also hinted that an intermediary could play an integrating role in the mobile development industry [3,4,10,15]. The more surprising observation is the fact that mainly phone manufacturer companies and software development companies have played this integration role and not so much MNOs as was the intuition of most of these scholars.

4   Implications for Developers

Hereafter we analyse the implication for developers of the three market trends presented in the previous section. In fact, the centralization of portal changes the way developers can distribute their application and reach a mass-market of consumers. The technological openness implies that developers would use different standards to develop their application and somehow work in a more collaborative mode. Then, highly-integrated platforms offer more possibilities to develop more sophisticated applications and services. These trends can be seen as opportunities but also threats for developers. Therefore, it is crucial that developers have a good understanding of the possible implications of each trend. They need to be able to choose the platform for which they want to develop knowing all the implications.

4.1 Implications of Portal Centralization

Portal centralization is an major shift for developers. It allows them to reach all potential customers through one shop, which takes care of the administrative tasks, such as billing and advertising. On top of these deployment facilities comes the fact that platforms providing centralized portals count on application sales to increase their revenue and therefore heavily promote application downloads and thus widely increasing the pool of potential consumers. This promotion is mostly done through advertising, but more importantly through greatly enhanced user interfaces. Before the emergence of centralized portals it took a expert user to download and install third-party applications, usually involving an internet search and a credit card payment, on a personal computer and then a file transfer via Bluetooth. Now it has become a "one-click" operation directly executable on the mobile device. Moreover, platforms can leverage on user communities which also promote applications using the reviewing features of the shops. A negative side of strong centralization for developers is that they might have to conform to certain rules defined by the portal provider. This problem can be observed with Apple's AppStore, which rules over which applications will be sold and which will be banned based on non-transparent criteria. To overcome these restrictions, the developer community has built alternative portals (Installer, Cydia) where developers can publish their applications. Unfortunately, only tech-savvy customers shop on such black markets, since phones must undergo a "jailbreak" procedure before they can access them.

4.2 Implications of Technological Openness

A move towards open source software offers two kinds of opportunities for application developers. First, as mentioned previously, moving towards open technology allows platform providers to reduce development costs and possibly increase the number of consumers. A greater number of platform consumers implies a greater number of potential application consumers for developers. Second, an open source project can provide career opportunities for developers willing to contribute to the platform development [8,9,13].

4.3 Implications of Platform Integration

The emergence of fully integrated end-to-end ecosystems, where the same people sell applications, manufacture devices and create their operating system, creates a coherent end-to-end approach, which makes it easier for applications to be developed, published, purchased, and used. There is less compatibility issues, which is a major problem in heterogeneous systems, where applications have to be fine-tune for specific devices with different display size for example. A drawback of high integration is the lack of alternatives if the solutions proposed by the platform do not suit the developer. Furthermore, higher integration means less need for platform interoperability, which implies that developers have to take sides and choose their preferred platform between the different contenders.

4.4   Implication of the Platform Choice

Choosing between platforms is not trivial for independent developers. We propose three criteria besides the personal identification with the platform, which plays a key role for some developers [8].

Income. First, the number of potential consumers who can be reached. A central aspect of the three previously mentionned trends is the fact that they increase the number of potential consumers through a mechanism such as lower prices, increase of usability, and better service. A higher number of consumers has the consequence of making the platform more appealing for developers, which will produce consequently more applications, which will make the platform more attractive for consumers. This creates a positive feedback loop, also called *two-sided network effects* [14,11]. Over the last year Apple has exhibited the strongest increase in consumers and mobile application developers, whereas Nokia still has the largest pool of potential consumers but fails to attract developers. Even though Apple is behind in the total number of consumers, it has the advantage of providing a centralized portal which facilitates consumer access.

Career. A second criterion could be the career opportunities that application development can lead to, i.e., being hired by a platform provider. As indicated above, open source development allows any developers to start working on the platform and possibly reach the committer level and then be hired by the company. With proprietary platforms only employees have access to the code therefore developers must first be hired. To increase their chances become recognized in the community, developers should join a young open source project [13]. Nokia and Google offer such an environment and the opportunities linked with it.

Freedom. Third, creative freedom is important to freelance developers. They must feel that they can program what they want. A well-prepared software development kit and an attractive mobile device in terms of features and performance can really generate interest and enthusiasm among independent developers. However, too many restrictions from the platforms can also produce negative effects. Therefore, open source platforms tend to provide more development freedom. LiMo and Google offer the best alternatives according to this criterion.

5   Conclusion

In this paper, we described the implications that different market and technology trends have on the mobile application development market. The current evolutions show that the game for the developers has changed dramatically. There are many new opportunities for them to develop, distribute, and generate significant revenues with the emerging mobile application portals. Since the mobile application development landscape has substantially changed over the past several years, mobile development platforms have become more integrated and generally play the role of application portal, device manufacturer or both. As discussed in the paper, application portals tend to become more centralized, facilitating the link between developers and consumers. Moreover, several new platforms entered the open source community to lower their costs and possibly extend their consumer market by lowering prices and as a consequence increase their developer pool. In this changing environment, choosing for which platform to develop reveals to be challenging and we proposed three simple criteria: market size and accessibility, career opportunities, and creative freedom.

References

1. Adrian, B.: Overview of the mobile payments market 2002 - 2007. Gartner (2002)
2. Ballon, P., Walravens, N., Spedalieri, A., Venezia, C.: The Reconfiguration of Mobile Service Provision: Towards Platform Business Models. In: Proceedings of ICIN 2008 (2008)
3. Stuart, J.B.: The mobile commerce value chain: analysis and future developments. International Journal of Information Management 22(2), 91–108 (2002)
4. Buellingen, F., Woerter, M.: Development perspectives, firm strategies and applications in mobile commerce. Journal of Business Research, Mobility and Markets: Emerging Outlines of M-Commerce 57(12), 1402–1408 (2004)
5. Capiluppi, A., Michlmayr, M.: From the cathedral to the bazaar: An empirical study of the lifecycle of volunteer community projects. Open Source Development, Adoption and Innovation, 31–42 (2007)
6. Funk, J.L.: The emerging value network in the mobile phone industry: The case of japan and its implications for the rest of the world. Telecommunications Policy 33(1-2), 4–18 (2009)
7. Gereffi, G., Humphrey, J., Sturgeon, T.: The governance of global value chains. Fortune Magazine (2005)
8. Hertel, G., Niedner, S., Herrmann, S.: Motivation of software developers in open source projects: an internet-based survey of contributors to the linux kernel. Research Policy 32(7), 1159–1177 (2003)
9. Lakhani, K.R., von Hippel, E.: How open source software works: 'free' user-to-user assistance. Research Policy 32(6), 923–943 (2003)
10. Maitland, C.F., Bauer, J.M., Westerveld, R.: The european market for mobile data: evolving value chains and industry structures. Telecommunications Policy 26(9-10), 485–504 (2002)
11. Parker, G.G.: Two-sided network effects: A theory of information product design. Management Science 51(10), 1494–1504 (2005)
12. Raymond, E.: The cathedral and the bazaar. Knowledge, Technology, and Policy 12, 23–49 (1999)
13. Riehle, D.: The economic motivation of open source software: Stakeholder perspectives. IEEE Computer, Article 25, 40(4), 25–32 (2007)
14. Shapiro, C., Varian, H.R.: Information Rules: A Strategic Guide to the Network Economy. Harvard Business School Press (November 1998)
15. Tsalgatidou, A., Pitoura, E.: Business models and transactions in mobile electronic commerce: requirements and properties. Computer Networks, Electronic Business Systems 37(2), 221–236 (2001)

A Middleware Architecture Supporting Native Mobile Agents for Wireless Sensor Networks

Ciarán Lynch and Dirk Pesch

Centre for Adaptive Wireless Systems,
Cork Institute of Technology,
Cork, Ireland
{ciaran.lynch,dirk.pesch}@cit.ie

Abstract. Mobile Software Agents are widely used in telecommunication networks and the Internet, however their application to embedded systems such as Wireless Sensor Networks is immature. We present a novel middleware supporting and enabling Mobile Agent applications to run natively, without any translation layer, on Wireless Sensor Networks. We establish that Mobile Agent systems are beneficial for a wide range of applications – particularly when dealing with complex, dynamic and spatially distributed tasks, and demonstrate their power and certain performance metrics for an example applications. We use an accurate emulation platform to evaluate the system performance in a distributed control application implemented using mobile software agents.

1 Introduction

The challenges of reprogramming Wireless Sensor Networks (WSN) after they have been deployed into the environment are well established [1, 2]. Reprogramming an entire sensor node program image is a time- and power-consuming process. Resetting the node also destroys any program state that has been established at the node.

Mobile agent systems have been proposed for WSN, however they either do not execute directly on sensor nodes or use an interpretation layer on the nodes. The system presented here uses native mobile agents, written in standard C, that execute directly on the embedded wireless sensor nodes.

Currently, there are no systems supporting true native mobile software agents on wireless sensor networks (see Related Work, Section 5). This paper presents a possible design of such a system. The proposed system builds on top of SOS [3], an existing modular operating system for WSN. Module support is tightly integrated into SOS and we found it a good base on which to develop such a system. The primary contribution of this paper is the development and evaluation of such a native mobile agent middleware system.

In order for such a system to be useful, it must minimise the cost of operation – primarily the time and energy used to transmit the agent code. The middleware system provides simple routing, reliable migration and remote module fetching,

C. Hesselman and C. Giannelli (Eds.): Mobilware 2009 Workshops, LNICST 12, pp. 65–74, 2009.

neighbour discovery and inter-agent communication. A weak mobility model is supported.

The architecture of the system is discussed in Section 2, and a simple application to evaluate it in Section 3. Section 4 discusses the results presented in the previous section. Related work is presented in Section 5, while Section 6 concludes the paper.

2   System Architecture

The Mobile Agent System Architecture is shown in Figure 1. It is split into a number of sub-sections. Agents execute as SOS modules, interacting with the middleware when they require the services it provides, running on standard MicaZ sensor nodes.

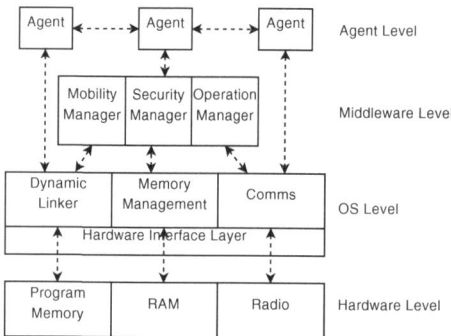

Fig. 1. System Block Diagram

The middleware must provide three types of services – Mobility, Security and Operations. The system is split into these three, essentially independent sections. The mobile agents as well as the middleware components are treated as normal code modules in SOS, with access to all of the operating system functionality.

2.1   Mobility Manager

An agent requires the ability to move. In the simplest case, an agent can move to another node within radio range. More complex movement operations enable more dynamic and powerful agents.

2.1.1   Agent Transfer

The manager controls the transmission of agents to neighbours. This can be either a move or a copy. The agent is suspended in order to capture a consistent snapshot of the agent state. An agent to be moved remains suspended until the transmission completes – if successful, the agent is then terminated, if not it resumes with an error signal.

A request is sent to the transfer node, and if it accepted, the transfer starts. The executable code is transmitted if necessary, followed by the agent state, including any dynamic memory allocated by that agent.

Once the transfer has started, the sending node sends packets regularly. Each packet carries 64 bytes of information. Selective-Repeat ARQ is used to ensure that all packets are received successfully. Since the size of an agent is generally limited to 10–20 packets, a sliding window would not be efficient. SR-ARQ gives better performance than simple ARQ in a lossy link, without overly complex resource requirements at the receiver.

2.1.2   Conditional Move

An agent may specify a move condition, allowing the agent to move to a point of interest in a network and not a neighbour. Intermediate nodes will still receive a copy of the agent, however they will not excecute it unless it has reached the end of the carrier path (the code is stored at the node for potential execution in the future). The condition has two parts, a move condition and a termination condition. The move condition is followed, until the termination condition is satisfied, or no further movement is possible.

Four carrier move conditions are currently defined – these are sufficient for a simple application, more may be added if required. They are *Move towards node N*, *Move along a gradient of a blackboard value* (see Section 2.3.1), *Move towards a gateway node* and *Move to a random neighbour*.

Five termination conditions are defined – *Execute at node N*, *Execute at the top or bottom of gradient*, *Execute at non-zero blackboard value*, *Execute at gateway node* and *Move once and then execute*.

2.2   Security Manager

A node should not blindly trust every agent received. Malicious users can inject improper agents, and errors or interference can result in an agent transmission becoming corrupted. A native agent does not have the protection of a bounded, virtual execution environment, and a corrupted agent will most likely crash the node. An MD5 signature [4] is used to sign each agent with a secret key, known only to the nodes in the system and to authorised users of the system.

2.2.1   Secret Key

Central to the signature scheme is a secret key known to each sensor node and to authorised users. This key is assumed to have been placed into each sensor node at the time of deployment or through a separate key distribution protocol.

2.2.2   Agent Integrity

The transmission fails if the hash does not match that generated locally. The sending or requesting agent is notified that the transmission has failed due to a security failure, and it is then free to take whatever subsequent action it deems appropriate.

The hash also operates as an integrity check – a corrupted agent that is not detected by the per-packet CRC check, or any misconfiguration in the middleware that causes the received image to be incomplete or incorrect will be rejected.

2.3   Operation Manager

2.3.1   Blackboard

The blackboard is a remotely accessible address space (similiar to Agilla Tuples [5]). A 16-bit key is used to index the space. The blackboard provides a simple, standard mechanism for discontinuous communication between agents in a local one-hop network neighbourhood. It allows agents to deposit information to be used in the future by other agents.

2.3.2   Neighbour Management

The manager maintains a list of neighbours. Agents can retrieve the list of current neighbours, query for new neighbours and select a random neighbour. Nodes are automatically removed from the list after a certain time.

2.3.3   Remote Agent Request

An agent requests the manager to find and execute another agent. If the agent code is available at the node, it is started. If not, a request for that agent is broadcast. Neighbouring nodes will forward this request and respond if the service is available. The agent is transferred by the mobility manager.

3   Evaluation

The middleware system was implemented and tested on MicaZ wireless sensor nodes. The application evaluation presented here is based on the AvroraZ emulation platform, as this provides detailed probes and instrumentation for measurement and testing, and allows repeatable and controllable experiments to be performed.

3.1   AvroraZ

AvroraZ [6] is an extension of Avrora, a wireless sensor network emulator. It uses a detailed model of the AtMega128L microcontroller to execute the same binary image as the physical microcontroller. It reproduces precisely the constraints of an embedded wireless sensor platform, allowing very accurate emulation of complex applications, complete with the timing and memory constraints of the physical system.

AvroraZ adds support for the CC2420 radio used in the MicaZ wireless sensor node platform. The radio model used is based on actual measurements of the performance of MicaZ nodes, and has been shown to correspond well with measurements from real sensor nodes [6].

3.2   Physical Scenario

The simulated scenario is a distributed control application. 25 nodes are placed in 3 rooms, with 3 nodes connected to heaters. The system is self-configuring, self-managing and robust – once one agent is pushed into the network, it operates autonomously, even in the presence of node resets and failures.

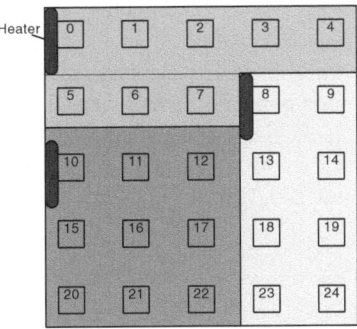

Fig. 2. Physical Scenario

The nodes are arranged as shown in Figure 2. Each node knows which room it is in (from 1 to 3) and whether or not it is attached to a heater. Every node has a temperature sensor. Each heater can be in one of 4 states – Off, Low, Medium or High. Each simulation is run for 6 minutes in total.

3.3   Agents

The agent sizes are shown in Table 4.

3.3.1   Management Agent

The management agent moves around the network, discovering nodes as it moves. On discovering a node connected to a heater, it summons a control agent if none exists. The control agent is given the list of nodes in the current room. This is updated each time the management agent revisits the control agent. The management agent keeps a list of nodes that were previously reachable but that are currently unreachable.

3.3.2   Control Agent (Static)

Every 10 seconds, it summons a data gathering agent to take readings from around the room. Once started, the control agent expects to get regular visits from the management agent. If one minute passes without any visit, the control agent will summon a new management agent, which will begin rediscovering the network from the current node.

3.3.3   Data Gather Agent

This agent moves around each node in the room and takes a sensor reading. It returns these readings to the control agent.

3.3.4   Reporting Agent

The reporting agent remains at an externally connected node. It receives the list of failed nodes from the management agent. Only a test agent is implemented.

3.4   Agent Middleware

The agent middleware, including the complete SOS operating system but without any agents occupies 77.8kB of the 128kB FLASH memory on the MicaZ node. This leaves 50.2kB for agent code. This is comfortably more than has been required in any application tested.

3.3kB of the 4kB of available RAM is used, including 2kB reserved for the dynamic memory heap – this is available to be used by agents and applications. The remaining 705 bytes is used for the program stack.

3.5   Results

Results of the evaluation are shown in Table 1. *Startup Time* is the time for the management agent to discover and visit every node in the system. *Management Period* is the average time between visits of the management agent to each node. *Gathered Values* is the average number of nodes visited by the data gather agent. Ideally this would be 8.3, however only 6.6 values are in fact being used per reading. Some moves fail due to congestion or due to incomplete network discovery at that point.

The performance of the middleware itself is given in Table 3. *Move Time* is the average total time for an agent move between nodes (per hop), including routing and transfer. *Routing Time* is the average time to find a particular node. The average time spent executing at each node for the datagather and management agents are shown as *Execution Time DG* and *Execution Time Man.*

The node setup was altered for some simulation runs, to test the resilience and self-management capability of the system (Table 2). Nodes 6, 17 and 19 are not started until 3, 4 and 5 minutes respectively. The time taken to incorporate them into the system is *Late Start*. *Reset Test* switches off nodes 9 and 10 after 140 and 250 seconds respectively, and reset and switches them on 50 seconds later. The time taken to reincorporate them into the system is almost the same as the late start test, as the mechanism is very similiar.

Table 1. Normal Operation

Operation	Value
Startup Time	58.12s
Management Period	18.72s
Gathered Values	6.59

Table 2. Failure Recovery

Operation	Value
Late Start	15.3s
Reset Test	15.9s

Table 3. Middleware Performance

Operation	Value
Move Time (per hop)	364.1ms
Routing Time	44.3ms
Execution Time (DG)	8.1ms
Execution Time (Man)	2.2ms

Table 4. Agent Sizes

Agent	Size (bytes)
Management	1220
Control	644
Data Gather	880

4   Significance and Discussion

4.1   Robustness and Self-management

The system uses interlocking agents to guarantee reliability. The loss of any agent or node in the network will not lead to complete system failure. If a node resets, the next time the management agent visits, the control agent will be reestablished. A new data gathering agent is generated for each operation, loss will simply limit the current control operation. The most damaging failure is the failure of the management agent. Assuming that at least one of the control agents survives, it will summon a new management agent, which will rediscover the network.

Once all of the agents have been established, the system is entirely self-managing. New nodes can be introduced into the network by simply placing them within the transmission radius of existing nodes and node failures are detected and resolved. Cooperating systems of mobile agents bring a new level of robustness to mobile code systems. A suitably designed system can work around or recover from complete node failures without any intervention from outside the network.

4.2   Structure and Flexibility

The primary benefit of the system is its flexibility. The dynamic distributed control application trades off node coverage and availability for the most up-to-date sensor data, requiring an autonomous and self-managing system. This tradeoff is a design decision and not a limitation of the middleware. The application designer is free to position the application at any point within this tradeoff, and has complete control over the performance.

The agent carrier system allows multihop transfers to take place seamlessly, and takes common mobility operations out of the agent, while the blackboard system allows inter-agent communication.

The native mobile software agent introduces a new level of performance for mobile applications, beyond what is currently possible with existing mobile agent systems for wireless sensor networks. This allows true autonomous multi-agent systems. It is only with networks of cooperating agents that the true benefits of mobile software agents are realised.

4.3   Alternatives

Applications such as those presented here, with dynamic discovery and network analysis, fault monitoring and control decisions are simply too complex to be implemented in a VM. The agent size benefit of a VM-based system is only realised the first time an agent visits each node – after this, only the operating state is transferred. The actual control operation could be implemented using TinyDB or some similiar data gathering tool. This requires an external agent to generate the queries and interpret the results – the sensor network is simply functioning as a data gathering tool.

5   Related Work

Existing mobile and dynamic code mechanisms for wireless sensor networks can be broadly separated at the highest level into application and presentation middleware [7] (Figure 3). Application middleware assists the application in executing on the wireless sensor network. They can be either native or interpreted. Native applications execute directly on the hardware of the wireless sensor nodes, while interpreted applications go through an interpretation or translation layer. Presentation middleware presents the network as a whole to the user, abstracting away the details of the hardware platform and topology. The work presented here is an example of an application middleware.

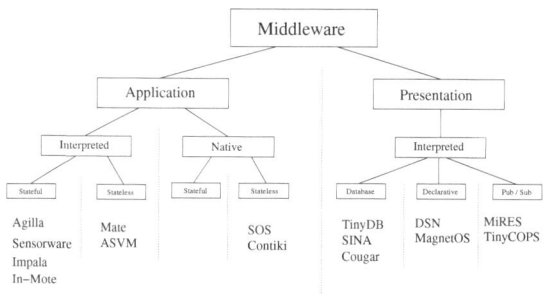

Fig. 3. Classification of existing middleware

All of these middlewares are capable of moving application code around the network. They can be divided into stateful middlewares, which transfer the application state, and stateless middlewares which do not.

5.1   Stateless Interpreted

ASVM [8] is a mobile agent systems for WSNs using a virtual machine (VM) to execute the agent code. ASVM allows extension and application customisation of the byte codes used in the VM, however the agent is constrained by the relatively simple VM in which it executes. TinyOS is required to execute the VM itself. Stateless systems essentially function as dynamic code distribution protocols. The work described here also contains such functionality but extends previous work by introducing stateful operation.

5.2   Stateless Native

Stateless native distribution systems are built into the various WSN operating systems, to send code updates through the network. Our middleware system is based on SOS [3] a dynamic operating system for WSN. This is the most suitable operating system at this time, however the middleware architecture is portable to other similar operating systems.

Contiki [9] also supports dynamic loading of modules, however much of the kernel is still compiled into a large static image. Much of the development of Contiki is focused on providing a large, highly capable networking stack.

The functionality of such distribution systems is duplicated in our work. We build on this by also providing state transfer and selective transfers, advancing the programming model from a simple static module to mobile and dynamic agents.

5.3   Stateful Interpreted

A stateful middleware is a form of mobile software agent [10]. Existing mobile agent systems for WSNs are either implemented on more capable devices such as PDAs or based on a Virtual Machine architecture. Impala [11] and Sensorware [12] are implemented on a PDA, written in Agent TCL. They provide support for complex dynamic applications.Scripts are injected into the network and can migrate, along with their state, to neighbouring nodes.

Agilla [5] is a more sophisticated version of ASVM, simpler than Sensorware but capable of operating on sensor nodes. Agents can move, copy and terminate themselves. Information is propagated among agents and one-hop neighbours using a tuple-space [13]. Nodes must have knowledge of their physical location and addresses are a function of the predetermined network topology.

The work described in this paper draws on much of the functionality of Agilla, but extends it to direct execution on the sensor node. The blackboard system is based on Agilla tuples, and the the simple agent mobility functions among one-hop neighbours are similiar. Our work however is more flexible and powerful in its execution model, and incorporates more powerful agent conditional transfers.

6   Conclusion

We have presented a native mobile software agent system for wireless sensor networks. It allows a level of complexity and performance in mobile software agents above that of existing systems. While the application presented here is only a preliminary examination of the system, it demonstrates its power and flexibility. Many design choices are available to the application designer, and the implementation can be tailored to suit the needs of the designer, rather than being constrained by the system in which it is implemented.

This flexibility does not come for free – agents must be carefully designed and the system will suffer in terms of application size. The safety of a virtual machine is not available, and agents must carefully manage their memory and resource usage. Simple applications will probably benefit little from moving to a native mobile agent system, however more complex applications can benefit greatly from the power and flexibility of the system, and the robustness and self-management that it achieves.

References

1. Jong, J., Culler, D.: Incremental network programming for wireless sensors. In: Proceedings of the First IEEE International Conference on Sensor and Ad Hoc Communications and Networks (October 2004)
2. Koshy, J., Pandey, R.: Remote incremental linking for energy-efficient reprogramming of sensor networks. In: Proceedings of the second European Workshop on Sensor Networks (EWSN 2005), Istanbul, Turkey (January 2005)
3. Han, C.-C., Kumar, R., Shea, R., Kohler, E., Srivastava, M.: A dynamic operating system for sensor nodes. In: Proceedings of the 3rd international Conference on Mobile Systems, Applications, and Services MobiSys 2005, Seattle, Washington, June 06 - 08, 2005, pp. 163–176. ACM Press, New York (2005)
4. Rivest, R.: The MD5 message-digest algorithm, Internet Engineering Task Force (IETF) RFC 1321 (1992)
5. Fok, C.-L., Roman, G.-C., Lu, C.: Rapid development and flexible deployment of adaptive wireless sensor network applications. In: Proceedings of the 24th International Conference on Distributed Computing Systems (ICDCS 2005), Columbus, Ohio, June 6-10, pp. 653–662 (2005)
6. de Paz Alberola, R., Pesch, D.: AvroraZ: Extending Avrora with an IEEE 802.15.4 compliant radio chip model. In: Proc. 3rd ACM International Workshop on Performance Monitoring, Measurement, and Evaluation of Heterogeneous Wireless and Wired Networks, Vancouver, BC, Canada (October 2008)
7. Sugihara, R., Gupta, R.K.: Programming models for sensor networks: A survey. ACM Transactions on Sensor Networks 4(2) (March 2008)
8. Levis, P., Gay, D., Culler, D.: Active sensor networks. In: Proceedings of the Second USENIX/ACM Symposium on Networked Systems Design and Implementation (NSDI 2005), May 2–4 (2005)
9. Dunkels, A., Grönvall, B., Voigt, T.: Contiki – a lightweight and flexible operating system for tiny networked sensors. In: Proceedings of the First IEEE Workshop on Embedded Networked Sensors 2004 (IEEE EmNetS-I), Tampa, Florida, USA (November 2004)
10. Tripath, A.R., Ahmeda, T., Karnik, N.M.: Experiences and future challenges in mobile agent programming. Microprocessors and Microsystems 25(2), 121–129 (2001)
11. Liu, T., Martonosi, M.: Impala: a middleware system for managing autonomic, parallel sensor systems. In: PPoPP 2003: Proceedings of the ninth ACM SIGPLAN symposium on Principles and practice of parallel programming, pp. 107–118. ACM Press, New York (2003)
12. Boulis, A., Han, C.-C., Srivastava, M.B.: Design and implementation of a framework for efficient and programmable sensor networks. In: MobiSys 2003: Proceedings of the 1st international conference on Mobile systems, applications and services, pp. 187–200. ACM Press, New York (2003)
13. Cabri, G., Leonardi, L., Zambonelli, F.: Mobile-agent coordination models for internet applications. IEEE Computer 33(2), 82–89 (2000)

Map-Based Compressive Sensing Model
for Wireless Sensor Network Architecture,
A Starting Point

Mohammadreza Mahmudimanesh, Abdelmajid Khelil[*], and Nasser Yazdani

University of Tehran, Technical University of Darmstadt, University of Tehran
m.mahmoudi@ece.ut.ac.ir, khelil@informatik.tu-darmstadt.de,
yazdani@ut.ac.ir

Abstract. Sub-Nyquist sampling techniques for Wireless Sensor Networks (WSN) are gaining increasing attention as an alternative method to capture natural events with desired quality while minimizing the number of active sensor nodes. Among those techniques, Compressive Sensing (CS) approaches are of special interest, because of their mathematically concrete foundations and efficient implementations. We describe how the geometrical representation of the sampling problem can influence the effectiveness and efficiency of CS algorithms. In this paper we introduce a Map-based model which exploits redundancy attributes of signals recorded from natural events to achieve an optimal representation of the signal.

Keywords: Wireless Sensor Networks, Compressive Sensing, Map-based WSN, WSN Architecture.

1 Introduction

A Wireless Sensor Network (WSN) is an instance of a distributed sensing network with a field of sensor nodes. But a very important constraint, differentiates a WSN from other similar configurations of sensing networks. WSN nodes usually have a very limited source of power. Every node must try to conserve as much energy as it can to extend the whole lifetime of the WSN. This energy is mostly consumed during sensing and multihop transport of the sensed data to a base station called the sink [2]. During the operation of a WSN usually a low quality of information (QoI) [11] is requested by the WSN user to make a decision about the occurrence of a specific event. Therefore, it is wiser to keep down as many nodes as possible to conserve more energy while satisfying user's needs [3]. The main responsibility of a WSN is to monitor the physical parameters of a natural operational environment (such as a desert and jungle). One attribute of signal representation of natural events (such as environment temperature and humidity) is that they mostly have very smooth changes and gradients over a plane. Compressive Sensing (CS) [4-7] is a novel sampling approach which tries to exploit compressibility of signals in order to reduce the

---

[*] Research supported in part by DFG GRK 1362 (TUD GKmM).

C. Hesselman and C. Giannelli (Eds.): Mobilware 2009 Workshops, LNICST 12, pp. 75–84, 2009.
© ICST Institute for Computer Sciences, Social-Informatics and Telecommunications Engineering 2009

minimum samples required to reconstruct the whole signal. CS offers a mathematically concrete method to capture only m samples of all n available samples of a signal (where $m \ll n$) and then exactly recover the original signal with an overwhelming probability. It exploits the compressibility attribute of a signal to wisely select only those samples which are important in signal reconstruction. We argue that based on the nature of environment and the geometrical distribution of the specific signal, we can find an arrangement that maximizes CS recovery performance and hence improves its quality.

The remainder of this paper is organized as follows. Section 2 gives a general view of our system model that we use throughout this paper. In Section 3 we summarize the mathematical basics of CS in general with special attention to its applications in WSN. In Section 4, based on our formulations in Sections 3, we discuss how selecting a suitable geometrical framework can improve the performance of compressive wireless sampling recovery. Section 4 presents an evaluation of the model, and Section 5 refers to some related works. Finally, we conclude with a summary and a brief description of future works.

2   System Model

Our WSN scenario consists of a network that is composed of n stationary resource-constraint sensor nodes (SN) and a static resource-rich sink. Commonly, WSNs are built utilizing hundreds to thousands of cheap SNs. For a WSN consisting of n SNs, there are $n!$ possible vectors of size n which can describe the environment as reported by sensors, at any instance of time. Here we focus on the spatial sampling of the environment using the SNs at a fixed instance of time. The main idea of our proposed model is to find a mapping between SNs and sensed vector elements that results the most compressible signal vector. We will formally define *compressibility* in the next section, but informally one can think of a compressible vector, as a vector that may have redundant data. The n SNs sample the physical signal f of interest, which we assume to be compressible. A data sample is characterized by the ID and the location of the SN as well as the sensor reading and its timestamp. We assume that SNs know their own geographic position. The clocks of SNs are synchronized, e.g., via GPS or any efficient synchronization protocol [12]. Positioning information is required to find the optimal spatial sampling method.

3   Compressive Sensing: Mathematical Basics and Applicability to WSN

Suppose a discrete signal f of n samples, which can be mathematically represented as a discrete-time n-dimensional vector signal $f \in \mathbb{R}^n$. CS is interested in *low-rate sampling*, in which the number m of available measurements is smaller than the dimension n of signal f. The vector f is in fact a discretely sampled version of a continuous signal of real environment. The sampling rate is either dictated by Nyquist sampling rate, or desired reconstruction resolution or quality. Therefore, when

$m < n$, we are indeed faced to a sub-Nyquist sampling situation. Reconstructing the actual signal $f \in \mathbb{R}^n$ from its sub-sampled version $y \in \mathbb{R}^m$ may seem infeasible. CS promises that if the signal satisfies some preconditions, it can be accurately or even exactly recovered from fewer *compressed samples*.

Definition 1: Sensing basis Φ is the basis in which signal is discretely represented. The classical case is when the sampling basis is delta Dirac waveforms, $\varphi_k(t) = \delta(t-k)$ that results in signal vector y consisting of samples $y_k = \langle f, \varphi_k \rangle$, $k = 1,\ldots,n$.

Definition 2: Sensing matrix is an $m \times n$ matrix A which is used to select an under-sampled edition of signal $f \in \mathbb{R}^n$, namely $y \in \mathbb{R}^n$, by the matrix multiplication: $y = Af$.

3.1 Sparse and Compressible Signals

Suppose we have a vector $f \in \mathbb{R}^n$ which we expand it in an orthonormal basis (such as a wavelet or Fourier basis) as follows:

$$f(t) = \sum_{k=1}^{n} x_k \psi_k(t) \tag{1}$$

where x is the coefficients vector under Ψ-transform of f, $x_k = \langle f, \psi_k \rangle$. More conveniently we can express f as Ψx, where Ψ is the $n \times n$ transformation matrix with $[\psi_1, \psi_2, \ldots, \psi_n]$ as columns. For example if we take Fourier basis as Ψ-basis, then $\psi_k(t) = n^{-1/2} e^{i2k\pi t/n}$ and x is the Fourier transform of f. Note that because we deal with time-discrete signals, t is limited only to integral values between 1 and n, $t \in \{1,\ldots,n\}$. Therefore, $\psi_k(t)$ determines the n items of the k-th column of matrix Ψ, by substituting t by $1, 2, \ldots, n$.

Definition 3: Representation basis Ψ is the basis in which the signal is transformed and represented for final storage or communication purposes. It can also be referred as Ψ-domain. The final purpose, is demanded by the specific application of signal sampling, compression, storage and recovery. For example wavelet-domain is a suitable Ψ-domain for image compression.

Definition 4: S-Sparse vector is a vector with only S nonzero items. We also call a signal f, S-Sparse, if its representation (in Ψ) is a S-Sparse vector.

Definition 5: Compressible Signal is a signal whose representation vector has many small (near zero) items and only a few relatively large and meaningful items.

In general for a vector v of size n, we define v_S as a S-sparse version of v by setting $n - S$ items of v to zero. By selecting a well-chosen S, the S-sparse vector f_S can be extracted from compressible signal f while maintaining the reconstruction error bellow a certain level. Hereafter, we may use "sparsity" and "compressibility" interchangeably. In the ongoing sections, sometimes by "sparsity" we mean that many elements of vector representation of signal are so small (near zero compared to other elements of vector), that we can easily neglect them.

3.2 Incoherent Sparse Sampling and Recovery

Suppose we are given a pair (Φ, Ψ) of orthonormal bases for vectors in $\mathbb{R}^n$. The first basis Φ is used for sampling the signal f in time or space domain, and the second is used to represent f in frequency domain. We define coherence between these two bases as follows:

Definition 6: Coherence. The coherence between the sensing basis Φ and the representation (frequency) basis Ψ is defined as:

$$\mu\ (\Phi, \Psi) = \sqrt{n}\ \max\left(\langle\varphi_k, \psi_j\rangle\right) \quad \text{over all} \quad 1 \le k, j \le n \tag{2}$$

in other words the coherence represents the largest correlation value between any two elements of Φ and Ψ.

CS is mainly concerned with low coherence pairs of sampling and representation bases. If Φ is the canonical or spike basis of delta Dirac functions ($\varphi_k(t) = \delta(t - k)$), and Ψ is the Fourier basis ($\psi_k(t) = n^{-1/2}e^{i2k\pi t/n}$), then it can be shown that $\mu(\Phi, \Psi) = 1$ and we have maximal incoherence. The interesting part of CS theory is that if we even select an orthobasis Φ uniformly at random, then with high probability, the coherence between Φ and Ψ is about $\sqrt{2 \log n}$. Random waveforms with independent identically distributed (i.i.d.) entries, also have a very low coherence with any fixed orthonormal representation bases Ψ.

Subsampling refers to sampling less than all available measurement. In WSN, this means that among all available n sensor nodes, we only observe a subset of all nodes and collect the data $y_k = \langle f, \varphi_k\rangle$, $k \in M$ where $M \subset \{1, \dots, n\}$ is a subset of cardinality $m < n$. Note that our reduced sensing basis has fewer dimensions, i.e. the signal is projected over fewer basis vectors.

Recovering original signal from these incomplete set of samples is performed by l_1-norm minimization [10]; the proposed reconstruction f^* is given by $f^* = \Psi x^*$, where x^* is the solution to the following convex optimization program:

$$\text{minimize} \|\tilde{x}\| \quad \text{subject to} \quad y_k = \langle \varphi_k, \Psi \tilde{x}\rangle, \quad \forall k \in M \tag{3}$$

where $\tilde{x} \in \mathbb{R}^n$ and l_1-norm is defined as $\|x\|_{l_1} := \Sigma_i |x_i|$. In simple words, among all vectors in Ψ-domain which are consistent with the collected data, select the x^* whose

l_1 norm is globally minimum. Then the recovered signal f^* can be calculated from Ψx^*.

Fundamental theorem of CS: Suppose that the Ψ transform x of f in the Ψ-domain is S-sparse. If we select m measurements in the Φ-domain uniformly at random so that

$$m \geq C \cdot \mu^2(\Phi, \Psi) \cdot S \cdot \log n \tag{4}$$

for some positive constant C, then the solution to the optimization problem (3) can accurately or even exactly recover the original signal f.

3.3 CS Advantages and Its Application in WSN

The following results from above mathematical summary leads us to think of CS as a very interesting and useful tool for sub-Nyquist sensor sampling, specially for WSN sensor reporting:

1) The smaller the coherence, the fewer samples are needed, hence CS emphasizes on low coherence systems. The measurement matrix can be even random or noise-like, because any randomly generated orthonormal basis has low coherence with Ψ transformation matrices such as Fourier or wavelet.
2) The resulting undersampled signal suffers almost no information loss if only about any random set of m coefficients are captured. The number of captured samples m may be far less than the signal dimensionality, if the coherence between sampling and representation bases is a small bounded value.

The CS theorem suggests a concrete and extremely efficient acquisition protocol: first sample the signal in an incoherent domain. Incoherence is the only perquisite; one can simply use random sampling. The sampling matrix needs not to be adaptive to the signal, i.e. this random subsampling technique can apply to any network topology and natural event recording. This makes CS to stand out as one of the best proposed sub-Nyquist sampling techniques for WSN. But as we see in the next section, we can even better utilize CS by considering geometrical attributes of the WSN environment.

4 CS-Oriented Map-Based (CSM) Architecture

Mapping is the core of the CSM model that reorders the vector signal elements in a way that maximizes CS recovery performance. In other words, mapping is the preparatory phase prior to CS sampling and recovery that prepares an alternative representation of sampling and recovery problem. We define mapping in an abstract way that assigns each element of signal vector f one and only one sensed value of the set of SNs readings. In comparison to other Map-based models for WSNs, our model is more abstract and general [9].

Definition 7: Mapping function M is a one-to-one function from $\{1,2,...,n\}$ to $\{1,2,...,n\}$ that defines a mapping between sensor readings and signal vector elements:

$$\forall k \in \{1,...,n\}, \quad f_k = s_{M(k)} \tag{5}$$

in which s is the vector of sensor readings arranged by SN identifier (ID) indexes.

As an example, suppose that s is the sensor readings vector ordered by sensor IDs of 3 sensors: (sensor 1, sensed value 13), (sensor 2, sensed value 11), (sensor 3, sensed value 10). Then $s = (13,11,10)$. Now we define the mapping function M as $M(1) = 2$, $M(2) = 1$, $M(3) = 3$. The resulting signal vector f under mapping M is $f = (s_2, s_1, s_3) = (11,13,10)$.

Mapping gives us the flexibility to choose a more efficient view of the environment. To show that why this flexibility is required, first we need to analyze the fundamental theorem of CS. For a fixed setup of a WSN in CSM model with n sensor nodes that uses Φ and Ψ as its sampling and representation bases, we find that $\mu^2(\Phi,\Psi)$ and $\log n$ are constants. Then (4) is reduced to the following linear equation:

$$m \geq C_0 S \tag{6}$$

where $C_0 = C \cdot \mu^2(\Phi,\Psi) \cdot \log n$. Then when all other parameters are fixed (which is of course so, after WSN is deployed), the minimum number of required random samples to recover the vector f depends only on the sparsity of the Ψ-transform of f. Note that the final recovery is done after applying the inverse mapping M^{-1} to vector f and assigning the recovered values to corresponding SNs.

In this model, a centralized or distributed processing entity is responsible for finding the best representation of the world with the most benefit. By representation we mean the mapping M, and benefit is in fact the sparsity (or compressibility) of vector f under mapping M. Here we don't try to present an algorithm to find the most beneficial mapping. Proposing an efficient algorithm can be another challenging topic, and can be even specialized for specific operational environments. In this paper we only try to show that WSNs that use CS approaches and exploit the geometrical properties of their operational environment can reach better CS recovery performance. This emphasizes the importance of considering Map-based WSN models in compressive wireless sensing (CWS) [1]. Next section presents a simple evaluation that runs a trivial exhaustive "efficient mapping" search algorithm to find the optimal mapping.

5   Evaluation of CSM with an Exhaustive Algorithm

Figures 1.a and 2.a show the actual environment in which our test sensor network will operate. In our tests, the environment is the same, but in each test we have randomly distributed the SNs over the operational area. The WSN consists of SNs that capture the temperature at different nodes of a natural area – such as a forest. In practice, such an image doesn't exist prior to sampling. In fact this figure is the ideal target of our WSN. We are going to apply CS sampling and recovery algorithms to this sensor network.

Our CS sampling method, selects randomly m of n sensor readings. Therefore we can even imagine of other $n-m$ sensors, being inactive. This is the situation that may happen in the case of sleep-scheduling energy preservation [2]. In fact the measurement matrix used in our test, was a simple $m \times n$ 0-1 matrix, that was constructed by randomly selecting m rows of $I_{n \times n}$. The Ψ-domain is the classical Fourier-domain, i.e. matrix Ψ is the inverse discrete Fourier transform matrix (IDFT). We have measured the quality of recovered vector, by computing the mean square error (MSE) between recovered and original vector. Because of the random nature of the tests, we ran each test for every WSN consisting of randomly distributed SNs, for one hundred times, and then averaged MSE of each run over all 100 results.

We evaluated and compared two sampling approaches:

a) In the first method we sample the nodes in a raster-like fashion from top to bottom and from left to right.

b) In the second method, we use the knowledge about the current actual temperature map. We examine SNs reported values and try to arrange them in a vector that is most compressible. This requires that for any possible mapping M, we reorder the sensor readings according to M, and compute the sparsity of its Ψ transform. The optimal mapping M^* is the one that generates the most compressible representation of vector f.

We don't try to invent an efficient algorithm that finds the optimal mapping in reasonable amount of time. We used a trivial exhaustive algorithm that searches among all $n!$ possible mappings and chooses the most optimal one. Because of the exhaustive search algorithm inefficiency, we tried the evaluation for a sensor network of only 10 SNs – that is $n = 10$.

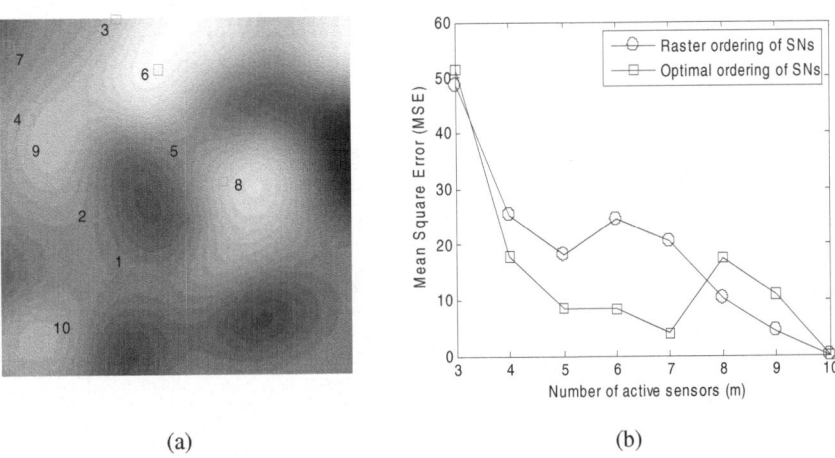

(a)   (b)

Fig. 1. a) Randomly distributed SNs of the first test. b) Average MSE for different cardinalities of randomly selected active sensor nodes.

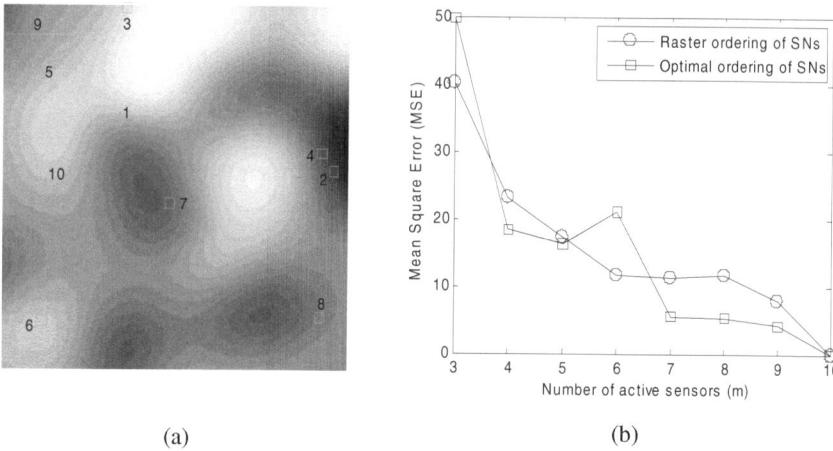

(a)   (b)

Fig. 2. a) Randomly distributed SNs of the second test. b) Average MSE for different cardinalities of randomly selected active sensor nodes.

Figures 1 and 2 depict two runs of our tests for two different positioning of SNs. In each of the figure pairs, Figure a shows the specific positioning of SNs in the test run, and Figure b shows a chart with the corresponding test results. In each of these two sample charts, the dashed line with circle markings determines the average MSE over all 100 CS recoveries of randomly sampled data using first sampling approach (a). Because of random nature of CS sampling, we perform the recovery procedure, one hundred times for each m in every test, and then take the average MSE as the actual MSE of that test. The dashed line with square markings shows averaged MSE using second sampling approach (b). For each simulated WSN, we run CS sampling and recovery for different number of active SNs. Because for this case, $m < 3$ can not give us a meaningful result, we started m from 3 to $n = 10$. As expected, by increasing the number of active SNs, MSE decreases and we have more accurate recovered signal vectors.

As we were expecting, the sampling strategy that tries to achieve the most compressible vector representation of environment signal, leads to lower MSE. However there are some anomalies for some m's, but in most cases, MSE of the tests based on sampling method (b) is lower than that of (a). The anomalies may occur because of randomness nature of problem, or small number of SNs. More satisfying tests can be performed using more SNs. But as we tested for some other testing WSNs, a similar behavior has been observed.

6 Related Work

There is a growing set of literature about specific uses of CS in WSNs, which try to model the WSN to fit the CS framework. Among them Compressive Wireless Sensing (CWS) [1] can be assumed to be pioneering work. Some other early papers have adapted CS for WSN and proposed concrete and efficient CS models for WSNs [13,14]. CWS methods are also tightly related to data compression techniques in

WSN [15-18]. But CWS has more practical results, as it promises efficient data acquisition protocol for distributed sensor networks with high cost of sampling.

Map-based World Model (MWM) for WSN refers to a generalized framework of WSN modelling, that views the world model of a WSN beyond a simple distributed sensing system [8]. As an improvement to previously proposed WSN design models, it also considers the topology of WSN nodes and specific geometrical distribution of the desired signal. Map-based models can be well adopted to exploit this attribute of natural event recording. By defining a suitable mapping that gives us a vector of signal samples with sparsest frequency-domain representation, we can improve the performance of CS signal reconstruction. The mapping procedure, that we have proposed in this work, is rather abstract and more general than MWM, but follows similar ideas.

There are also another set of proposed data acquisition techniques which specially try to exploit the spatial correlation of SNs reportings [19-21]. One can put all series of work under a more general field of study, that deals with distributed sampling in a relatively error-prone and resource-constrained network architecture. In such situations, because of some information loss, we try to reconstruct the signal from a fewer number of available samples [22-24].

7 Conclusion and Future Work

In this paper, we have introduced a new approach to signal acquisition in WSN. We have stated that selecting a suitable geometric framework can utilize better the smoothness attribute of natural signals. By exploiting the spacial attributes of the operational environment, a more compressible vector view of environment signal can be represented. This affects an essential factor of Compressive Sensing (CS) techniques performance, called the *sparsity* of signal in *representation basis*. The sparser or more compressible the representation of a signal, the fewer samples needed to be captured for signal reconstruction using CS algorithms.

However, we didn't try to derive an efficient algorithm to find the optimal mapping, but presented a test on a small scale WSN, which shows that the approach that follows our CS-oriented Map-based (CSM) model, can achieve better results. The lack of an efficient optimal map finding algorithm, limits us to try the model on a WSN consisting of more SNs. Deriving such an efficient algorithm can be a future challenge. But for this small-scale testing WSN, number of tries are high enough to fade away the effect of random sampling.

References

1. Bajwa, W., Haupt, J., Sayeed, A., Nowak, R.: Compressive wireless sensing. In: Int. Conf. on Information Processing in Sensor Networks, IPSN (2006)
2. Akyildiz, I.F., Su, W., Sankarasubramaniam, Y., Cayirci, E.: Wireless sensor networks: a survey. Comput. Networks 38 (2002)
3. Raghavendra, C.S., Sivalingam, K.M., Znati, T. (eds.): Wireless Sensor Networks, 2nd edn. (2005) ISBN: 978-1-4020-7883-5

4. Candès, E., Wakin, M.: An introduction to compressive sampling. IEEE Signal Processing Magazine 25(2) (2008)
5. Candès, E.: Compressive sampling. Int. Congress of Mathematics, Madrid, Spain (2006)
6. Baraniuk, R.: Compressive sensing. IEEE Signal Processing Magazine 24(4) (2007)
7. Donoho, D.: Compressed sensing. IEEE Trans. on Information Theory 52(4) (2006)
8. Khelil, A., Shaikh, F.K., Ayari, B., Suri, N.: MWM: A Map-based World Model for Event-driven Wireless Sensor Networks. In: Proc. of The 2nd ACM International Conference on Autonomic Computing and Communication Systems, AUTONOMICS (2008)
9. Khelil, A., Shaikh, F.K., Ali, A., Suri, N.: gMAP: An Efficient Construction of Global Maps for Mobility- Assisted Wireless Sensor Networks. In: The Sixth Annual Conference on Wireless On demand Network Systems and Services, WONS (2009)
10. Candès, E., Romberg, J.: Sparsity and incoherence in compressive sampling. Inverse Problems 23(3) (2007) ISSN 0266-5611
11. Zahedi, S., Bisdikian, C.: A framework for QoI-inspired analysis for sensor network deployment planning. In: 2nd Int'l Workshop on Performance Control in Wireless Sensor Networks, PWSN (2007)
12. Sundararaman, B., et al.: Clock Synchronization for Wireless Sensor Networks: A Survey. Ad-Hoc Networks 3(3) (May 2005)
13. Cevher, V., Gurbuz, A.C., McClellan, J.H., Chellappa, R.: Compressive wireless arrays for bearing estimation of sparse sources in angle domain. In: ICASSP 2008 (2008)
14. Hern, B.: Robustness of Compressed Sensing in Sensor Networks, Bachelore thesis (2008)
15. Kimura, N., Latifi, S.: A Survey on Data Compression in Wireless Sensor Networks. In: Proceedings of the international Conference on information Technology: Coding and Computing (ITCC), vol. II, pp. 8–13. IEEE Computer Society, Los Alamitos (2005)
16. Barr, K., Asanovi, K.: Energy aware lossless data compression. In: Proceedings of the 1st international Conference on Mobile Systems, Applications and Services (MobiSys), pp. 231–244. ACM Press, New York (2003)
17. Kusuma, J., Doherty, L., Ramchandran, K.: Distributed compression for sensor networks. In: Proc. International Conf. Image Processing (ICIP), October 2001, vol. 1, pp. 82–85 (2001)
18. Arici, T., Gedik, B., Altunbasak, Y., Liu, L.: PINCO: a Pipelined In-Network Compression Scheme for Data Collection in Wireless Sensor Networks. In: Proceedings of 12th International Conference on Computer Communications and Networks (October 2003)
19. Adler, M.: Collecting correlated information from a sensor network. In: Proceedings of the ACM-SIAM Symposium on Discrete Algorithms, SODA (2005)
20. Chu, D., Deshpande, A., Hellerstein, J., Hong, W.: Approximate data collection in sensor networks using probabilistic models. In: Proceedings of the International Conference on Data Engineering, ICDE (2006)
21. Pattem, S., Krishnamachari, B., Govindan, R.: The impact of spatial correlation on routing with compression in wireless sensor networks. In: Proceedings of the International Conference on Information Processing in Sensor Networks, IPSN (2004)
22. Pradhan, S., Ramchandran, K.: Distributed source coding using syndromes (DISCUS): Design and construction. IEEE Transactions on Information Theory 49(3) (2003)
23. Silberstein, A., Puggioni, G., Gelfand, A., Munagala, K., Yang, J.: Making Sense of Suppressions and Failures in Sensor Data: A Bayesian Approach. In: Proceedings of the International Conference on Very Large Data Bases, VLDB (2007)
24. Slepian, D., Wolf, J.: Noiseless coding of correlated information sources. IEEE Transactions on Information Theory 19(4) (1973)

Mobility and Remote-Code Execution

Eric Sanchis

University of Toulouse 1,
IUT, 33 avenue du 8 mai, 12000 Rodez, France
sanchis@iut-rodez.fr

Abstract. Using an adapted analysis grid, this paper presents a new reading of the concepts underlying the *mobile code/agent* technology by proposing a decomposition of the paradigms related to *remote-code execution* into three categories: *remote-code calling*, *remote code-loading* and *mobile code*. Models resulting from this decomposition are specified and implemented using a uniform execution system. A distinction between *mobile code* and *mobile software agent* is then proposed.

Keywords: distributed systems, design abstraction, remote-code execution, mobile code, remote code-loading.

1 Introduction

Distributed applications implemented with mobile code or mobile agents were actually developed from the first half of the Nineties [1]. Mainly considered under a technical angle - mechanisms, programming -, opinions are divided today on the utility of mobile agents in these applications [2], [3], [4]. Indeed, as a mechanism of remote-code execution, *mobile code* is considered to be less universal than *remote procedure call* while being more difficult to implement. Moreover, no *killer application* truly emerged. Nevertheless, a certain number of implementations showed that *mobile code* was an interesting mechanism in perfectly targeted applications.

When we wish to study *mobile code* as a manner of executing remote code, we come up against the multiple meanings which are associated with the concept of *code* such as:

- An executable code C provided with its data D and its execution context E: the entity which moves is a complete execution unit defined by the triplet (C, D, E). This type of mobility is generally called *strong mobility*
- An autonomous code or a code fragment C provided with some initialization data D: the couple (C, D) is downloaded on the remote site then executed by a new execution unit (*weak mobility*)
- A non mobile procedure P provided with its parameters which first must be bound to an execution unit before being called (*remote procedure call*).

In these three cases, the concept of *code* has different semantic contents. Moreover, we can notice that neither *weak mobility*, nor *remote procedure call* need the *execution context* to be brought in.

C. Hesselman and C. Giannelli (Eds.): Mobilware 2009 Workshops, LNICST 12, pp. 85–97, 2009.
© ICST Institute for Computer Sciences, Social-Informatics and Telecommunications Engineering 2009

The various interpretations of the notion of *code* and the consideration or not of the *execution context* brought us to revisit the paradigms connected to the distributed execution domain and to re-evaluate the place that the *mobile code* holds in this area. Our work led us to the following conclusion: *weak mobility* and *strong mobility* belong to quite distinct design paradigms. *Weak mobility* just like *remote procedure call* aim at the execution of a remote code, while *strong mobility* is centred on the migration of an execution context. That means that *weak mobility* and *remote procedure call* correspond to the same abstraction that we have called *Remote-Code Execution (RCE)* and which will be detailed in the following sections, whereas *strong mobility* is conceptually close to the migration of process or thread, i.e. a paradigm which we could call *Migration of Execution Unit*. In other words, *weak mobility* is closer to *remote procedure call* than to *strong mobility* because an execution context cannot be reduced to its code part.

In order to position *mobile code* with regard to the architectures generally used within the distributed applications framework, a second reading of the paradigms related to *RCE* was carried out starting from the work and proposals presented in [5].

This paper is structured as follows. Section 2 explains in depth the grid of analysis which was used to carry out our second reading, a grid built on the concepts of *Abstraction, Model* and *Mechanism* (called *A2M* grid) where each level of the grid masks a set of non relevant details. Section 3 compares the Fuggetta's design paradigms and the models of our *RCE* abstraction. The last section describes the *execution system* which was used to implement the *RCE* models.

2   Abstraction, Model and Mechanism

The construction of an IT application requires the use of several classes of services provided by the underlying system (communication, synchronization, naming services). In order to extract the principles at the heart of a class of services and to ignore the implementation details, it is preferable to reason about an *abstraction* representing a class of services solving a standard system problem. For example, the conceptual or concrete functioning modalities relative to the services of communication between processes can be grouped together in an *interprocess communication* abstraction. However, it seems that the notion of communication even limited to simple processes takes on very different meanings according to distributed system designers. As far as we are concerned, we define a *communication* as the transmission of a sequence of bytes between a sender process and a local or remote receiver process. Other approaches utilize the concept of *object* or *exchange* (call/reply, transaction) [6], [7], [8]: it is not any more an elementary communication between two processes but a *structured communication* between active entities, with the various additional variations which it authorizes. Consequently, although dealing both with communications, they are two different abstractions. The essential interest of *abstractions* is that they present a global comprehensive view of a problem and their associated solutions (formalized or implemented) when this problem is perfectly limited.

Generally, for each *abstraction* one or more *models* are defined where each one of them formalizes a particular manner to solve the target problem. *A model* is characterized by a set of features, in general a few ones, which identify its specificity and which make it possible to immediately distinguish it from the other models of the same abstraction. Thus, the *interprocess communication* abstraction as we described previously declines according to the two well-known models: the *shared memory* model and the *message-passing* model.

Lastly, while a model shows certain coherence, it can be implemented using very different *mechanisms*. This multiple forms of the same model is due to several factors such as the programming language used or the characteristics of the chosen execution support (resources management, interactions between the execution units, etc.). Thus, inside the Unix family systems, several mechanisms implement the model of *communication by shared memory*. Let us quote for example the communication by *pipe, shared memory* or *message queue* (an inadequate denomination with regard to the implemented model!). The properties of these three mechanisms are very different with respect to the communication direction (one-way or bidirectional), to the synchronization of the communicating entities (synchronization provided by the execution support or by the programmer). For complex abstractions such as *RCE*, the associated mechanisms appear under the form of powerful execution systems which section 4 will give some examples of.

It should be noted that often, relations between a model and its implementations are sufficiently strong to introduce a certain *duality* between the model and mechanism concepts. Indeed, certain *mechanisms* studied in a given context, are presented as *models* in another context. For example, the *RPC mechanism* (*Remote Procedure Call*) is often promoted to the rank of model, generally called *Client-Server model*. We can attribute the emergence of this *model/mechanism duality* to the consideration of different elements by the designers to define their paradigms. That will be clarified in section 3 when the study of the paradigms related to the code mobility presented in [5] will be made, paradigms which will be compared with our own models associated to *RCE*.

Moreover, the *plurifunctionality* of certain models/mechanisms used in various contexts leads to a situation where there is no unanimity among the researchers to place a model in a single abstraction. For example, the *Client-Server* model is often considered as a model of communication between process or threads [9], [8], [10] and not as a model of the *Remote-Code Execution* abstraction.

Lastly, some complex mechanisms integrate sub-mechanisms which correspond to models belonging either to the same class, or to different classes. This aspect will be illustrated in section 4 where the *middleware* used to implement the *Mobile Code* model is built on the *Client-Server* model.

In order to dissipate as well as possible the negative effects carried by the *model/mechanism duality* and by the *plurifunctionality* previously described, we proposed an analysis grid *A2M* which is articulated around three adjacent conceptual levels (Figure 1), offering a certain flexibility in the definition and interpretation of the studied paradigms.

The porosity of these three levels already present in the communication between processes is more important when the studied abstraction is more complex such as *RCE*. In order to specify the contents of this abstraction, the following section presents the relationships between the *design paradigms* of a distributed system, *mobile code* and the *RCE* abstraction.

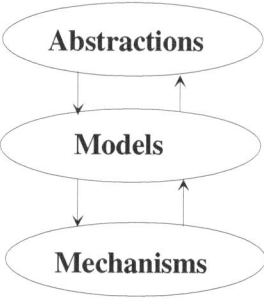

Fig. 1. The A2M grid

3 The Remote-Code Execution Abstraction

A. Fugetta and his colleagues [5] studied in depth paradigms connected to *mobile code*. Their project aimed at clarifying the terms, concepts and technologies which are related to these paradigms. Their work led to the definition of a three-dimension architecture: the *technologies*, the *design paradigms* and the *applications* of mobile code. With regard to these three axes, the second reading we propose focuses on the *design paradigms* (called *models* in the *A2M grid*). This is why only this aspect will be explained in detail thereafter.

3.1 Design Paradigms

These researchers identify four general design paradigms: *Client-Server (CS)*, *remote evaluation (REV)*, *code on demand (COD)* and *mobile agent (MA)*. In order to clarify the characteristics of these four paradigms, they distinguish the following elements:

- a component *A*, located on site *Sa*, which requests the execution of a service *S* and waits for the corresponding result
- a component *B*, localized on site *Sb*
- the requested service *S*
- the necessary resources *R* (data, files, etc.).

The four paradigms are then decomposed into two categories: *Client-Server* and those which exploit the code mobility (*remote evaluation*, *code on demand* and *mobile agent*).

Using the *Component/Service/Resource* triptych, the *Client Server* paradigm (CS) is stated as follows: at the initial moment, the service *S* and the resources *R* are localized on site *Sb*. The component *A* located on site *Sa* asks the component *B* to execute the service *S*. After execution of S, *B* returns the result to *A*.

The three paradigms associated to the code mobility are described in the following way.

Remote evaluation (REV): at the initial moment, component *A* located on site *Sa* possesses the required service *S* but the necessary resources *R* for obtaining the result are present on site *Sb*. Component *A* sends service *S* to component *B* located on *Sb*. *B* uses resources *R* to execute service *S*, then returns the result to *A*.

Code on demand (COD): at the initial moment, component *A* has the necessary resources *R* to execute service *S* but this one is located on site *Sb*. *A* interacts with component *B* which sends to it the required service *S*. *A* obtains the expected result by executing on its site the received service *S*.

Mobile agent (MA): at the initial moment, *A* possesses service *S* and a part of the resources *R* necessary to its execution, the other part of the resources being on the site *Sb*. After a partial execution of service *S* on site *Sa*, component *A* provided with service *S* moves on site *Sb* where it continues the execution of *S*.

As it was noticed by the authors, there is an unequivocal division between on one side the *REV* and *COD* paradigms, and on the other side the *MA* paradigm. This separation is attributed to the fact that in the *MA* paradigm, there is not only the movement of a service but the transfer of an execution unit. As we suggested in section 1, this formal asymmetry is due to the consideration of two design paradigms belonging to two different abstractions: the *RCE* and *Execution Unit Migration* abstractions.

Within the framework of a methodology based on the use of the *A2M* grid and before defining conceptually homogeneous design paradigms (models), it is necessary to carefully characterize the contents of the abstraction which will synthesize these models.

3.2 Models of the Remote-Code Execution Abstraction

Defining an *abstraction* consists in clarifying the generic problem, each model of the abstraction providing a specific resolution method. As its naming indicates, the aim of the *Remote-Code Execution* abstraction is the execution of a piece of code present on a remote host. That means that the characterization of the models of this abstraction will be based essentially on the *code* part, and more specifically on the operations on this code such as *copy*, *execution* and *deletion*. To illustrate in a uniform manner the functioning of the various *RCE* models and particularly the flow of control between the local active entity and the remote one, only *synchronous* interactions will be considered. That means that neither asynchronous alternatives, nor various types of *resources* (in all its forms: initialization data, execution contexts - stack, instruction pointer -, opened file descriptors, etc) will be taken into account in modelling.

A precision must be underlined about the models naming which will follow. As the description of the models is only based on the operations concerning the *code* part, the denomination of the models will be different from the names of the paradigms described in section 3.1. Thus, the *Client-Server* paradigm is named *Remote Code Calling* model in the *RCE* abstraction. This naming has two advantages: on the one hand, it does not use the *Client-Server* expression which is used in many branches of computing and on the other hand it specifically refers to the operation carried out on the *code*, i.e. the aspect which is at the heart of the model. The equivalents of *REV* and *COD* paradigms are considered as two versions of the same principle: the *Remote Code-Loading*. Finally, the *MA* paradigm disappears from *RCE* abstraction for the benefit of a new model: the *physically mobile code* (or more simply *mobile code*).

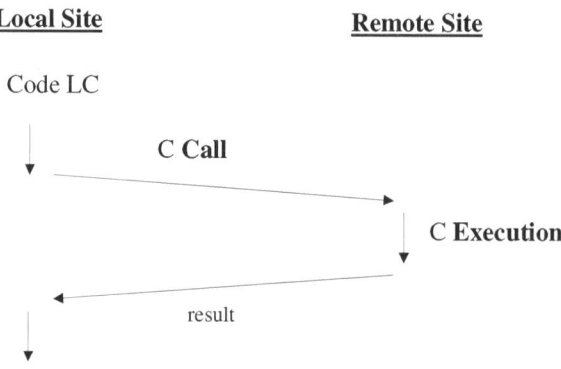

Fig. 2. Remote Code Calling

Remote Code Calling. Figure 2 illustrates the *Remote Code Calling* model.

This model generalizes the traditional *procedure call* mechanism in a distributed system. The principle of *Remote Code Calling* is the following: at the initial moment, the local code *LC* requires the execution of the remote code *C* present on site *Sb* and waits for the result. After execution of code *C* on *Sb*, the result is returned to code *LC* which continues its execution.

It is important to notice that no precision relative to the data, the used resources or the intermediate synchronizations between the execution units associated with codes *LC* and *C* are present into the model: these details are encapsulated into the model implementation according to the used *execution system*.

Remote Code-Loading. It is with the *Remote-Code Evaluation* and *Code on Demand* models that important differences appear with the *REV* and *COD* paradigms.

Remote-Code Evaluation is a generalization of the *evaluation* principle implemented by the interpreters of certain high-level languages. This model can be described as follows: at the initial moment, the local code *LC* copies a code *C* on site *Sb* and asks for its execution there. Code *C* is executed on *Sb*, the result is returned to

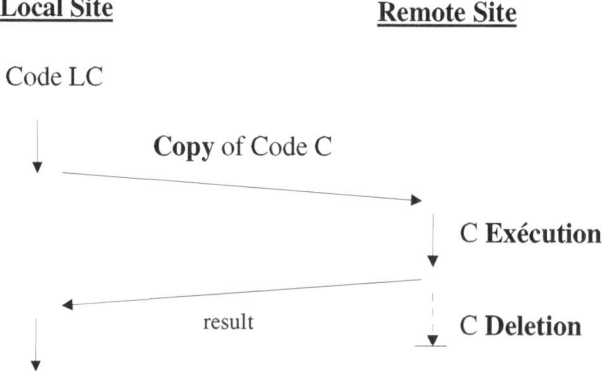

Fig. 3. Remote-Code Evaluation

code *LC* then code *C* is removed from *Sb* (Figure 3). After the result reception, code *LC* continues its execution.

It is the combination of the *Copy/Execution/Deletion* operations which characterizes the remote evaluation of code *C*. This model substitutes the code mobility aspect by a *remote code-loading* sketch. This point of view is perfectly compatible with the application which is usually used to illustrate the *REV* paradigm: the printing of a PostScript file. When a copy of this file is loaded, this copy is directly interpreted by the printer then disappears.

The same operations combination also applies to *Code on Demand*, but executed in an indirect way (Figure 4).

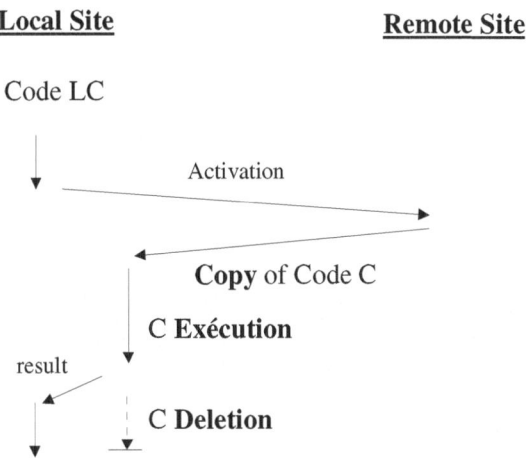

Fig. 4. Code on Demand

The principle of the *Code on Demand* model is the following: at the initial moment, the local code *LC* activates an intermediate remote code *RC* located on *Sb*. Code *RC* copies on *Sa* code *C* which is also located on *Sb*. Code *C* is executed on *Sa*, the result is provided to code *LC* and code *C* is removed from *Sa*. After the reception of the result, code *LC* continues its execution.

The presence of the code deletion in the *Remote-Code Evaluation* and *Code on Demand* models contributes to reinforce the internal coherence of the two models. Indeed, if the operation of deletion was absent, the successive execution of two remote evaluations of the same code *C* on the same remote site *Sb* would lead to a logically incoherent behaviour (copy of code already present).

Mobile code. The *Mobile Code* model derives directly from the *physical mobility* of an object, mobility understood according to the common sense: a mobile object is an object which is present in its physical totality at a point *x* at an instant *t*, then is at a point *y* (*y* different from *x*) at instant *t+1*. This model extends the concept of passive message (*short lifespan data*) to the concept of active message (*persistent lifespan code*).

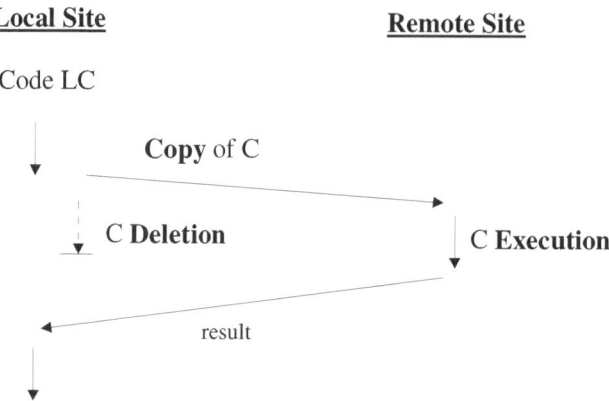

Fig. 5. Mobile Code

The *Mobile Code* principle is translated in the following way: at the initial moment, local code *LC* copies code *C* to site *Sb* then removes code *C* from *Sa*. Code *C* is executed on *Sb* and the result is returned to code *LC* which can continue its execution. Disappeared from site *Sa*, code *C* remains present on the remote location *Sb*: there was an actual physical moving of code between the two sites (Figure 5).

By construction, two successive executions of the same mobile code *C* cannot take place from the same site *Sa*. This behaviour corresponds perfectly to the natural semantics of a code qualified as mobile. The mobility of the code is actual.

Lastly, to be complete and by orthogonality with the *Code on Demand* model, we deduct the indirect version of *Mobile Code* (Figure 6).

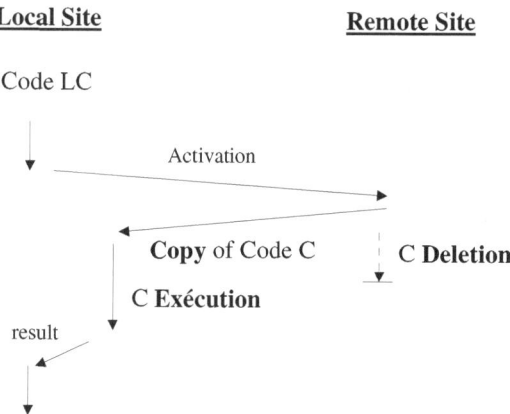

Fig. 6. Indirect Mobile Code

Compared to the *Remote-Code Evaluation* model, in these two models the operations of execution and deletion are made on two different sites.

3.3 Discussion

The *RCE* abstraction as previously characterized, i.e. centred on the code part and the associated operations, has the major consequence to completely reorganize paradigms that are generally related to *mobile code*. Indeed, we showed that:

- *remote evaluation* (paradigm *REV*) and *code on demand* (paradigm *COD*) are two models built on the remotely copying of code and not on its mobility: thus, they must be clearly distinguished from *mobile code*. Our point of view is that neither PostScript printing nor Applet and Servlet technologies constitute examples of *mobile code* but are different implementations of the *remote code-loading* model. On the other hand, advantages generally attributed to *mobile code* or *mobile agent* such as the reduction of the network load or the interest to bring closer code and data to be treated are not related to the mobility but to the copy of the code on the data's site. In other words, certain positive aspects associated to *mobile code* are rather to put at the credit of the *remote code-loading* model

- *strong mobility agents* do not constitute a model of the *RCE* abstraction but a model of a very different abstraction – *Execution Unit Migration* -, an abstraction the study of which (characterization of the abstraction, its models and mechanisms) remains to be made. Consequently, the advantages and disadvantages associated with *strong mobility agents* are not attributable to the management of their code but rather to the mobility of their execution unit as in *process migration* implemented in certain operating systems [11].

The *mobile code* model of the *RCE* abstraction completes in a coherent way the tools offered to the distributed applications designer, tools dedicated to remote-code execution. By construction, this model allows the checking of the uniqueness and the location of the executed code. It also illustrates the difference between *mobile code* and *strong mobility agent*.

Lastly, it invalidates the widespread idea according to which *viruses* and *worms* would be mobile entities. Indeed, the code of these software entities does not move from a site to another but remotely replicates itself on infected sites: if *viruses* and *worms* were mobile, it would be easier to eradicate them. Their power of nuisance results essentially from their capacity of remote replication.

4 From Models to RCE Mechanisms

The purpose of the models presented previously was to extract the specificity of each of them with regard to the other models of the *RCE* abstraction. This specificity was formalized by a particular combination of operations relating to the *code* part, one of the singularities of the suggested interpretation being to consider all other aspects as concerning a particular implementation. Before illustrating a building of each model, i.e. to pass from the model to a mechanism which implements it, the used execution support must first be described. Indeed, it is the characteristics of the underlying *execution system* that will determine the possible synchronizations, the necessary resources and the global modes of functioning.

The judicious choice of the used *execution system* associated with the formal coherence of the *RCE* abstraction models allowed us an implementation of these models which is precise, elegant and homogeneous as well.

4.1 Execution System

Two elements characterize an *execution system*: the *middleware* and the *programming language* which are used.

A *middleware* groups together a set of broad utility services making the implementation of distributed applications easier. Often, a full development system goes with the middleware, a system which not only provides the necessary tools for the construction of the distributed application, but the tools facilitating its deployment too. Two examples of very used middlewares are the RPC and RMI systems. The first one is intended for the implementation of distributed applications written in C language, the second to those written in Java. Both middlewares realize the same remote-code execution model (*Remote Code Calling*), but in different ways: *Remote Procedure Call* for the first and *Remote Method Invocation* for the second.

Two classes of programming languages are mainly used to implement a distributed application: *system programming languages* such as C, C++ or Java and *script languages* such as the Unix shells (bash, sh, ksh), PERL, TCL or Python. Although each class has its advantages and its disadvantages, *script languages* have a strong power of expression perfectly adapted to the use we wish to make [12].

For more simplicity, the *execution system* which was used to implement the various models of the *RCE* abstraction is articulated around the middleware **SSh** and the script language **bash**. The *execution system* **bash/SSh** was selected for the two main following reasons: a native deployment on many computing systems and a concise and elegant programming syntax.

Middleware **SSh** provides a high level network programming interface with the couple of secure commands **ssh/scp**. Well configured, the middleware **SSh** offers a sufficient security level to simply implement the *RCE* abstraction models. The remote-code execution model at the centre of the commands **ssh/scp** is *Remote Code Calling*. By default, interactions via **ssh** or **scp** between the two entities are synchronous. However, the interpreter **bash** natively integrates the necessary mechanisms to implement *asynchronous* interactions.

Shell **bash** - as any other Unix shell - interfaces itself with **ssh/scp** in a perfect manner. Its compactness favours the fast writing of prototypes. Weakly typified and string oriented, it is less sensitive to the traditional problems posed by data representation between heterogeneous systems than system programming languages such as C or other languages of the same family.

4.2 RCE Models Implementation

Supplying an implementation of the *RCE* models simply consists in translating the operations defined in section 3 into comprehensible instructions by the *execution system* **bash/SSh**. These instructions are the following:

To execute the command *cmd* located on the remote site *Sb*: **ssh Sb cmd**

To copy on remote site *Sb* the code of command *cmd*: **scp cmd Sb:**

To locally copy the code of command *cmd* located on remote site *Sb*:

      scp Sb:cmd   .

To remove the code of command *cmd*: **rm cmd**

To assign to a variable *var* the value resulting from the execution of command *cmd*:

      var=$(cmd)

The translation of the models is immediate (to simplify, the various cases of error are not treated).

Remote Code Calling.

```
result=$( ssh   Sb   C )
```

The result coming from the execution of the remote code *C* is assigned to the local variable *result*.

Remote Code-Loading.

Remote-Code Evaluation

```
scp   C   Sb:
result=$( ssh   Sb   "C   ;   rm C" )
```

Code *C* is copied to the remote site *Sb*. Then, the two commands *C* and *rm C* are sequentially executed on *Sb*. The result of the remote execution of *C* is assigned to the local variable *result*.

Code on Demand

```
scp Sb:C   .
result=$(C)
rm   C
```

Code *C* located on the remote site *Sb* is locally copied, locally executed then locally removed.

Mobile Code.

```
scp   C   Sb:
rm   C
result=$( ssh Sb C)
```

Code *C* is copied to the remote site *Sb* then the local copy of *C* is removed. Code *C* is executed on site *Sb* and the result is assigned to the local variable *result*.

The indirect version of *Mobile Code* is expressed in the following way:

```
scp Sb:C   .
ssh  Sb  rm C
result=$(C)
```

5 Conclusion

In previous works [13], the study of complex properties such as *autonomy* revealed how important it was to use a precise reasoning framework to be able to analyze then to rigorously classify the models associated with the same paradigm. It was shown that an incompletely controlled abstraction could lead to inappropriate conclusions.

The definition then the use of grid *A2M* applied to *mobility* showed the relevance to distinguish *mobile code* and *strong mobile agent*: the *mobile code* model belongs to the *Remote-Code Execution* abstraction while *strong mobility* is to be associated with the *Execution Unit Migration* abstraction.

The meticulous construction of the *RCE* models clarified the *replication/mobility* duality and underlined the important features making it possible to distinguish the various models of the *RCE* abstraction. The immediate profit was to restore to each model its advantages and its disadvantages.

References

1. Chess, D.M., Harrison, C.G., Kershenbaum, A.: Mobile Agents: Are they a good idea? IBM Research Report, RC 19887 (1994)
2. Lange, D.B., Oshima, M.: Seven Good Reasons for Mobile Agents. Communication of the ACM 42(3), 88–89 (1999)
3. Vigna, G.: Mobile Agents: Ten Reasons For Failure. In: Proceedings of the IEEE International Conference on Mobile Data Management 2004 (MDM 2004), Berkeley, USA, pp. 298–299 (2004)
4. Johansen, D.: Mobile Agents: Right Concept, Wrong Approach. In: Proceedings of the IEEE International Conference on Mobile Data Management 2004 (MDM 2004), Berkeley, USA (2004)
5. Fuggetta, A., Picco, G.P., Vigna, G.: Understanding Code mobility. IEEE Transactions on Software Engineering, 24(5), 352–361 (1998)
6. Goscinski, A.: Distributed Operating Systems – The Logical Design. Addison Wesley, Reading (1991)
7. Silcock, J., Goscinski, A.: Message Passing, Remote Procedure Calls and Distributed Shared Memory as Communication Paradigms for Distributed Systems. Technical Report TR C95/20, School of Computing and Mathematics, Deakin University (1995)
8. Coulouris, G., Dollimore, J., Kindberg, T.: Distributed Systems – Concepts and Design. Addison Wesley/Pearson Education (2005)
9. Tanenbaum, A.: Modern Operating Systems. Prentice Hall, Englewood Cliffs (1992)
10. Silberschatz, A., Galvin, P.B., Gagne, G.: Operating System Concepts with Java. John Wiley & Sons, Chichester (2007)

11. Thiel, G.: LOCUS operating system, a transparent system. Computer Communication 14(6), 336–346 (1991)
12. Ousterhout, J.K.: Scripting: Higher Level Programming for the 21st Century. IEEE Computer 31(3), 23–30 (1998)
13. Sanchis, E.: Autonomy with Regard to an Attribute. In: IEEE/WIC/ACM International Conference on Intelligent Agent Technology 2007 (IAT 2007), Silicon Valley, USA (2007)

A Component-Based Approach for Realizing User-Centric Adaptive Systems

Gilbert Beyer, Moritz Hammer, Christian Kroiss, and Andreas Schroeder

Ludwig-Maximilians-Universität München
{beyer,hammer,kroiss,schroeda}@pst.ifi.lmu.de

Abstract. We discuss a generic architecture for building user-centric systems. The characteristic feature of such systems is a control loop that monitors the user's state, and produces a harmonized response. In order to adaptively respond to changes of the user's state, we propose an architecture with supervising loops. This allows the primary control loop to be written in a straight-forward way, and add adaptivity on a different level. We illustrate our approach with an example scenario that describes computer vision based adaptive interaction.

Keywords: Components, reflective systems, adaptation, reconfiguration.

1 Introduction

Software and software-enhanced systems are becoming more and more interwoven with our everyday life. Likewise, it is well understood that software systems have to pervade and adapt autonomously to the user and the environment in order to increase their value to him [Mar99]. The idea that software systems also consider the "nonfunctional aspects" of the environment and especially the user, i.e. the emotional state (such as anger or joy), cognitive engagement (such as high attention or distraction) and physical conditions (such as body posture and fatigue), is however relatively new [PK01].

How to design and realize such human-centred pervasive adaptive applications is still hardly understood, as so far most approaches to adaptive systems have been concentrating on delivering "pure functionality" to the user, e.g. providing enhanced information through location-based services.

Realizing human-centred pervasive adaptive applications is challenging in two ways: first, the users requirements and intentions cannot be queried directly, but need to be inferred from unobtrusive, nondisruptive measurements of the user and his environment. Second, foreseeing possible adaptation dimensions of a pervasive application and realizing them in the software is a task requiring a careful and systematic design approach.

In this paper, we present the vision and concepts of the REFLECTive framework and middleware: a software framework that is intended to facilitate the construction of user-centric pervasive adaptive systems. In the next Section, we first clarify our vision on user-centric pervasive adaptation. Then, we present our

C. Hesselman and C. Giannelli (Eds.): Mobilware 2009 Workshops, LNICST 12, pp. 98–104, 2009.

software-engineering approach to the design of user-centric pervasive-adaptive software, allowing to create well-structured and flexible software applications in Section 3, and discuss their benefits in one example scenario in Section 4. Thereafter, we discuss related work in Section 5, and conclude in Section 6.

2   User-Centric Pervasive Adaptation

User-centric pervasive adaptive systems try to improve the well-being of a user by influencing environmental and workload-related parameters with the goal of improving the user's physical, emotional, and cognitive state. In order to infer the user's state, a user-centric pervasive adaptive system use available context information (e.g. is the user currently driving for hours in a traffic jam), and psychophysiological measures (e.g. skin conductance, heart rate, or EEG). This data allows to fine-tune the environment of the user to create a more comfortable setting.

However, realising such systems brings its own challenges. For example, it is never guaranteed that an action performed on the environment (such as e.g. changing the lighting conditions) will show the desired effects, or any effects at all: preferences vary greatly over persons and over time. User-centric pervasive adaptive systems will therefore have to self-asses their effects on their environment, and hold several alternative strategies ready to achieve their goal. Furthermore, extracting the user's conditions from low-level psycho-physiological measures or computing it through image processing is a challenging task with varying success depending on the user's context; detecting the facial expression under bad lighting conditions for example is very difficult, if not impossible at all. These challenges call for a well-structured software solution that allows for flexible response to the users environment, and self-assessment of its performance. We foresee that in order to tackle this application domain, the software system built adapting the users environment and thereby guiding the users conditions towards well being needs to be adaptive and flexible in itself.

As the ability for internal restructuring to our eyes is one key enabler for building functioning feedback loops, one of the challenges in our approach to user-centric pervasive adaptive systems therefore is how to enable and facilitate this internal adaptation; This question is answered in the following sections.

3   Realizing User-Centric Software

Our primary approach towards user-centering are control loops. Realizing such loops, however, is a challenging task of software engineering. Given the unpredictability and uncertainty involved when considering a human user, control loops need to be flexible and adaptive, possibly need to integrating multiple sensor readings and are subjected to long-term changes of the application's goals. In order to facilitate the development of such loops, we propose an architecture that tries to address the various concerns involved in control loops separately.

For a clean structure of control loops, we suggest to distinguish a number of phases that are executed in sequence within the loop. We adopt the MAPE scheme as proposed by IBM [KC03]. This scheme separates four distinct stages: Monitoring, Analysis, Planning and Execution. First, data needs to be obtained by monitoring. This data is analysed in the next stage, resulting in a concise description of the state. This description serves as input to the third stage, which is concerned with planning an appropriate action. Finally, the fourth stage executes the plan.

By separating the stages, a clear view on the various aspects to be considered when devising a control loop is obtained. In the context of user-centric systems, each stage poses its own challenges: monitoring needs to access physical sensors, possibly preparing the data, analysis needs to provide an abstract description of the user's state. Planning then needs to produce a plan that integrates the application's goals (e.g., mood improvement) and the observed user's state. This plan is to be executed in the last stage, using available actuators like displays or sound devices. Each of these stages is difficult to realize even without considering adaptivity. On the other hand, each stage is more or less required to be adaptive: the monitoring phase might have to address the loss of physical sensors, while the planning phase might need to respond to the failure of previous plans to achieve the intended effect.

In order to keep the realization of control loops tractable, we suggest to externalize the concern of adaptivity. Instead of putting adaptivity-related code into the MAPE loop stages, we suggest to supervise the MAPE loop, and replace some of its stages, should they prove inadequate. This externalization is obtained by two design elements: the use of *components* to build the control loops, and the use of *hierarchical* loops to obtain supervision.

Components have been among the first concepts provided for obtaining a modular design of software. Basically, they are black-box entities of software that explicitly declare their communication endpoints. In our approach, components declare a number of *ports*, which declare *provided* and *required* communication capabilities. Ports of different components can be connected by means of *connectors*. Since we need to handle a variety of communication styles (e.g., sensor readings which are transported from hardware layers to the monitoring components, or discrete messages that communicate the abstracted user state from the analysis stage to the planning components), we suggest to use *fat connectors*, which handle the implementation of the communication (as opposed to predefined communication means supported by the framework).

In order to facilitate clear application design, we also propose the use of hierarchical components. A component can then either be an atomic component, implemented in a host programming language, or a composite component that consists of a number of components and connectors that connect some of the ports of these components. The ports that are not connected within the hierarchical components become the ports the hierarchical components exports. Eventually, an entire application can be built as a hierarchical components that does not export any ports.

By building applications (or, for our purpose, stages of a MAPE loop) by selecting and connecting components, the architecture remains visible. It can then be changed at runtime, in a process called *reconfiguration*. Reconfiguration can replace components, add or remove filter components to connections, or just modify a component's parameters. We suggest to realize adaptivity in this way: instead of integrating different strategies within a single component, we propose that reconfiguration is used to obtain adaptivity. For example, if a sensor fails, the monitoring component (which should be a composite component, consisting of various components that obtain and process the data) should be reconfigured to continue without the sensor.

Doing reconfiguration is not trivial, however. For utilizing reconfiguration as a means to obtain adaptivity, the four stages of the MAPE loop need again to be implemented: first, the application that might become reconfigured needs to be observed. From this observations, an abstract state – and a decision about the necessity of reconfiguration – needs to be generated. If reconfiguration is deemed necessary, it has to be planned in order to obtain an architecture that is more suitable. Finally, the plan has to be executed in a safe way (i.e., the reconfiguration must not disrupt ongoing communication [KM90]).

We hence propose a higher-level MAPE loop that supervises the low-level control loop and changes it by the means of reconfiguration if the low-level control loop is not performing as intended. Obviously, the higher-level MAPE loop can again be subjected to supervision, achieving a way to change long-term goals of an application.

4   Example Scenario: Adaptive Interactive Installations

A possible field of application of our proposed architecture for adaptive systems can be found in interactive installations in public spaces. Examples of such user-centric media are interactive tables, displays or projections in museums, exhibitions or other public buildings, providing topic-oriented content when users interact with them. Interactive installations can realize both a 1-to-1 dialogue with the user as well as interaction with a group of visitors [JS04]. 1-to-1 dialogue means that the installation communicates with a single user or is explicitly controlled by him. Installations that communicate to many people are put into effect by multi-user environments that offer a shared experience. The requirements for both types of user interaction are typically different. For example it often doesn't make sense to let an interactive object or virtual world be controlled by more than one user the same time; an interactive game instead may require a minimum of participants to get evident. The application designer has to decide to comply with one scenario that best fits the presumed situation. Interactive installations that are able to adapt to variable user situations (cf. Fig. 1a and 1b) are an example of applications that could expediently make use of the component-based approach and hierarchic MAPE loops.

The hierarchy of MAPE loops in this scenario can be structured as follows (cf. Fig. 2): an inner loop would realize the preferred interaction (e.g. explicit control of content by a camera sensor and hand gestures) and update the displayed content accordingly. A higher-level MAPE loop could assess if the difficulty level

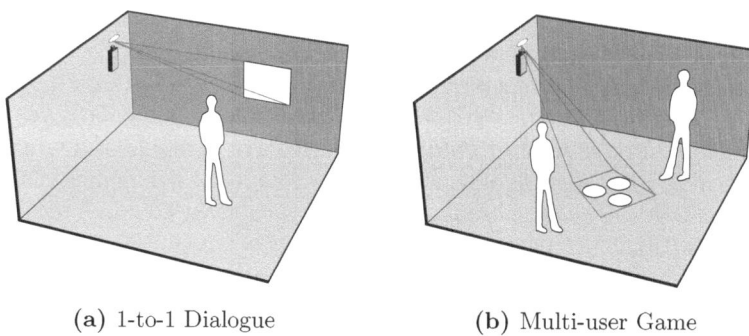

(a) 1-to-1 Dialogue   (b) Multi-user Game

Fig. 1. Adaptive Interactive Installation

for the user has to be adjusted (e.g. if the user is moving his hands too slow) and adapt the corresponding parameters of the underlying interaction control loop. At the third level, another MAPE loop could monitor the current situation of the system (e.g. the number of users within range). In this step, the system would assess if the requirements for the chosen interaction are still fullfilled or if the content type of the lowest MAPE loop has to be changed, e.g. from one that supports one-person to one that supports multi-user interaction. To realize an adaptation like this by parameter adjustment can be quite cumbersome. Instead, the involved component within the lowest MAPE loop could be replaced by others that realize the desired interaction control.

As a way to facilitate the creation of such pervasive adaptive systems, the Java-based REFLECT component framework [WBH+09] is developed. It is built

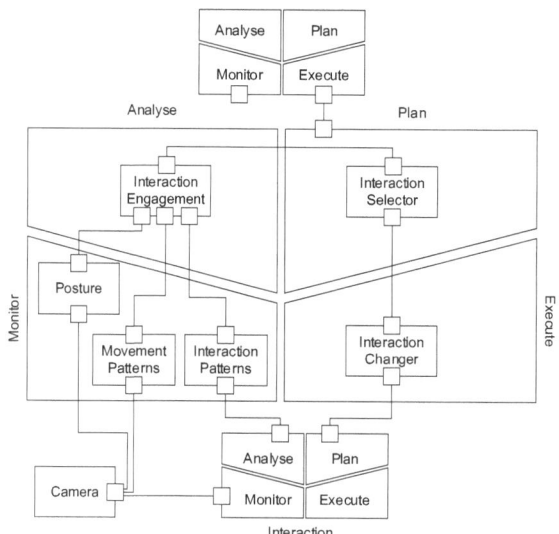

Fig. 2. Scenario Structure

on top of the OSGI service platform [OSG05] and provides a hierarchical component model with means for reconfiguration and reflection. Based on this foundation, some experimental applications were realized to evaluate the applicability of our approach and to test the framework. These applications instantiate the general scenario described above and focus on its technological challenges. The component-based approach and the concept of hierarchical and reconfigurable MAPE loops could be veryfied to be highly beneficial, especially as the involved computer vision techniques require the combined use of several complex algorithms. These could be very well structured as interlinked components that are adaptable by parameter-adjustment or reconfiguration. Additionally, rapid prototyping and experimentation were highly facilitated by assembling applications from reusable components. This is very important since user-centric adaptation constitutes a relatively new field of research.

5   Related Work

A large number of middlewares supporting adaptivity have been developed. A rather broad overview is given by Sadjadi et al. [MSKC04]. Huebscher and McCann [HM08] focus on the utilization of control loops for adaptivity, identifying an extended MAPE loop as a common denominator.

In [KM07], a three-layered architectural reference model for self-managed software is described that comprises layers for component control, change management and goal management, respectively. Reconfiguration is regarded as the essential adaptation mechanism of the component control layer. The change management layer contains a collection of change plans that are executed as reactions to state changes. The goal management layer is concerned with producing new change plans according to high-level goals. Altogether, this architecture could be interpreted as a hierarchy of MAPE loops with distinct responsibilities.

6   Conclusion and Future Work

We have presented an approach towards engineering user-centric systems. Since user-centric systems need to be constantly aware of the user's state, our approach builds on control loops. Implementing such loops is not easy, and we propose a two-fold separation of concerns for facilitating the development: considering discrete stages, and externalizing the adaptivity to a supervising loop. As part of the REFLECT project we have built a framework that supports this way of building user-sentric applications, and gathered first experience with interactive display systems.

For building larger applications, we consider further support for the MAPE loops to be necessary. This especially holds true for the supervising MAPE loops: analysing the need for a reconfiguration requires detailed knowledge about the purpose of the lower-level MAPE loop, and planning a reconfiguration needs additional knowledge about the alternative architectures. We believe that future work should address this issues, since automating parts of the MAPE loop appears a necessity for large-scale applications.

References

[HM08]   Huebscher, M.C., McCann, J.A.: A survey of autonomic computing—
            degrees, models, and applications. ACM Computing Surveys 40(3), 1–28
            (2008)
[JS04]   Sauter, J.: Das vierte format: Die fassade als mediale haut der architektur.
            In: Digitale Transformationen. Medienkunst als Schnittstelle von Kunst,
            Wissenschaft, Wirtschaft und Gesellschaft. whois verlags und vertriebs-
            gesellschaft (2004)
[KC03]   Kephart, J.O., Chess, D.M.: The vision of autonomic computing. Com-
            puter 36(1), 41–50 (2003)
[KM90]   Kramer, J., Magee, J.: The evolving philosophers problem: Dynamic
            change management. IEEE Transactions on Software Engineering 16(11),
            1293–1306 (1990)
[KM07]   Kramer, J., Magee, J.: Self-managed systems: an architectural challenge.
            In: Future of Software Engineering (FOSE 2007), pp. 259–268. IEEE Com-
            puter Society Press, Los Alamitos (2007)
[Mar99]   Mark, W.: The computer for the 21st century. SIGMOBILE Mob. Com-
            put. Commun. Rev. 3(3), 3–11 (1999)
[MSKC04]   McKinley, P.K., Sadjadi, S.M., Kasten, E.P., Cheng, B.H.C.: A taxonomy
            of compositional adaptation. Technical Report MSU-CSE-04-17, Depart-
            ment of Computer Science, Michigan State University (2004)
[OSG05]   The OSGi Alliance. OSGi service platform core specification release 4
            (2005), http://www.osgi.org/Release4
[PK01]   Picard, R.W., Klein, J.: Computers that recognise and respond to user
            emotion: theoretical and practical implications. Interacting with Comput-
            ers 14(2), 141–169 (2001)
[WBH+09]   Wirsing, M., Beyer, G., Hammer, M., Kroiss, C., Schroeder, A.: RE-
            FLECT - first year report: Requirements and design. Technical report,
            LMU München (2009)

A Reflective Goal-Based System for Context-Aware Adaptation

Dejian Meng and Stefan Poslad

School of Electronic Engineering and Computer Science, Queen Mary University of London,
Mile End Road, London E1 4NS, United Kindom
{dejian.meng,stefan.poslad}@elec.qmul.ac.uk

Abstract. Many context-aware applications exploit current world contexts to control information access and service adaptation. To support a user centric utility model for a context-aware system, user goals are supported by the system. However, traditional goal-based approaches such as planning are hard to achieve in a context-aware environment because environments and systems are dynamic. This can lead to inconsistencies and goal conflicts. In this research, a new goal-context based approach is introduced to guide service adaptation for context-aware applications. A reflection model is investigated and applied to the system to resolve ambiguities and inconsistencies between goal-based planning and actual service adaptation. The new system, with its improved flexibility and robustness, is demonstrated in the form of a mobile spatial routing application.

Keywords: Context-aware; Reflection; Adaptation; Goal-based; Spatial routing.

1 Introduction

Mobile applications [1] [2] [3] [4] can exploit spatial routing services to plan all kinds of traffic routes for mobile users. Mobile applications can configure or customize these routing services according to user needs. When the environment where the application is situated changes, the application will sometimes need to reconfigure these services to match different context conditions. Context-awareness, in this case is deployed to enable the mobile application to adapt its service configuration. In a user-centric pervasive environment, such as travel, the environment should include not only the physical environment such as current location, traffic flow, etc., but also the user environment. The user environment model can include a user's situation, tasks, preferences and goals.

To introduce support for user goals into the context-aware system, several problems need to be addressed: how user goals can be defined so that a continuous and variable user expectation or preferences to the application can be recognized; how a goal-based approach and dynamic planning can be used to introduce more flexibility into existing static context-service relationships; how goal-based planning can handle user situation, physical environment and system changes; How any ambiguities and inconsistencies in the planning process that occur can be handled. In this paper, a

C. Hesselman and C. Giannelli (Eds.): Mobilware 2009 Workshops, LNICST 12, pp. 105–110, 2009.
© ICST Institute for Computer Sciences, Social-Informatics and Telecommunications Engineering 2009

reflective user goal-based context-aware framework is proposed to address these design issues.

2   A Reflective User Goal-Based Context-Aware Framework

In this context-aware framework, context models especially the goal context model are defined. Two layers of adaptation, i.e., context adaptation and service adaptation are defined. The context adaptation layer enables planning to be deployed during context evolution (from current contexts to goal contexts), which provides a way to exploit goal contexts to reflect user expectations about the interaction. A service adaptation layer performs the actual configuration or customization of services. A reflective middleware between the two layers is used to resolve any inconsistencies.

2.1   Contexts

Context-aware systems are systems that are aware of the situation in their physical and virtual environment and can respond in some way to benefit from knowledge of that situation. A context, according to Dey and Abowd [5], can be defined as any information that can be used to characterize the situation of an entity, considered relevant to the interaction between a user and an application. Spatial routing applications require information such as current location, time, weather, traffic situation and so on to characterize route environment changes, so that it can be aware of their changes and adapt to them.

Context adaptation can be represented as a transformation from current user contexts to users' goal contexts. In the spatial routing application, users expect routes to be generated to satisfy all their constraints and preferences. The simple view is that the destination location is the goal context in the spatial routing example, because the application user is heading to the destination. However, in terms of the spatial routing application, a more complex view of the adaptation should also take into account goal constraints of the route as goal contexts, i.e., a route to the destination that is short, fast or scenic. The goal context might be changing by the time, as user expectations in such a dynamic environment will be different.

Two dimensions of a goal context are modeled: the intrinsic goal (path) context defines how the goal context is reached from a corresponding current context; while the extrinsic goal (constraints) context defines meta-dimension information about the goal dimension information, e.g., how the goal context is constrained. The difference relies on whether it is information about the goal (user expected) situation, or the goal (user expected) information about the current situation. The goal dimension configuration specifies rules to change the configuration of goal contexts, e.g., the weights of each goal context. Some concrete context instances for a spatial routing application are shown in Fig. 1. For example, if it is known that other users will arrive late at a meeting venue, a fast route becomes less important because the user is not in a hurry as his or her preferences for the goal constraints have changed.

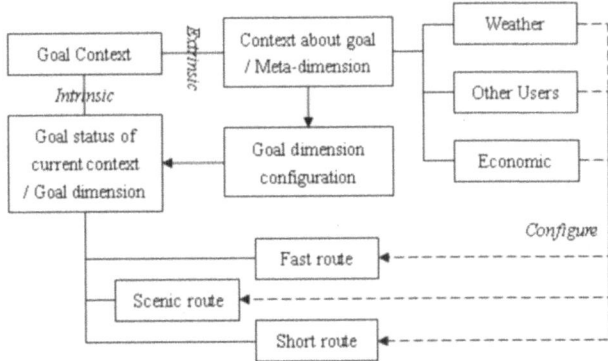

Fig. 1. The goal context model defined in terms of a goal path and goal constraints

2.2 Adaptation Layers

Service adaptation refers to the adaptation of service instances and to the dynamic customization of services. The adaptation of service instances can be multi-valued, involving reconfiguration of weights to determine how a goal path is generated according to the combination of one or multiple goal constraints e.g., for a spatial routing application [6], the weights of how fast, how scenic, and how short the route is can be adjusted. Applications execute the expectation of the user through instantiating service customization. External contexts drive service adaptation to change the current situation to the goal situation.

Most designs for context-aware computing use built-in mechanisms such as condition-action rules for service adaptation. There is a lack of a holistic approach to combine the use of multiple individual contexts to adapt to goal contexts, hindering more complex adaptation. This often involves deliberating about context changes rather than purely reacting to context changes. Static rules lack flexibility because the knowledge that supports decisions is not represented explicitly and cannot be modified. A layered adaptation model for a context-aware system is defined comprising a Context Adaptation Layer (CAL) and a Service Adaptation Layer (SAL). CAL is made up of a context model and a context path. In the CAL, current contexts and goal contexts need to be combined so as to establish an ordered (fully or partially) path to guide service adaptation in SAL. Context adaptation is a process which guides service adaptation. In this case, a context adaptation specification will be transformed into a service adaptation specification, which links (external) context adaptation to (internal system) service adaptation. Therefore an interface is defined between the external world (context) and internal world (service). A service adaptation layer actually executes the services which are driven by the external context adaptation layer (Fig. 2).

The multi-lateral context-aware adaptation architecture (Fig.2) is made up of a context adaptation layer, a service adaptation layer and middleware. Goal contexts are abstracted from the user expectation to the application. Planning is used to generate a context adaptation specification for a context path from the current to the goal

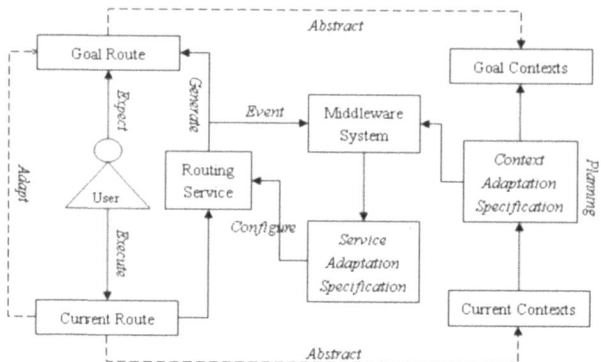

Fig. 2. A multi-lateral context-aware adaptation architecture

context. When a configured service does not generate a goal route that meets the goal constraints, the inconsistency is spotted and reported back to the middleware as an event to be resolved (see next section).

2.3   The Reflective Middleware

With loosely-coupled context and service models, inconsistencies between context changes and service changes can occur. For example, in spatial routing applications, if there is a traffic accident deviation to the planed route, CAL will normally plan to avoid a blocked road, minimizing the added deviation to satisfy a multi-valued goal context such as a particular shortest, fastest and scenic route. This leads to a routing failure in SAL because the traveling time from such a routing plan will exceed the time limits the user expected. To solve this problem, reflection about such potential failures takes place before the actual route is generated. To this end, Adaptation Middleware Layer (AML) also adds a reflective model about the internal system to the existing external environment model.

A reflective system focuses on meta-computation, computation to reason about its own operational computation [7]. A reflective system needs a representation of its own computation. The operational status and computation of the system complies with this representation [8]. The key to applying a reflective model to the existing system is that the self-representation can be modified at run-time and these modifications actually have an impact on the run-time computation. Regarding our system, we add AML to the existing spatial routing service which is located at the SAL to make the service adaptation reflective (Fig. 3).

Context composition combines different context types in a multi-valued context and instances of contexts. It uses a planner to link the current contexts and goal contexts. The planning process is achieved by partitioning the goal contexts further into sub-goal contexts and reasoning about the corresponding knowledge defined in the context domain to meet those sub-goal contexts. A context adaptation specification will be generated from the context composition, which will be mapped

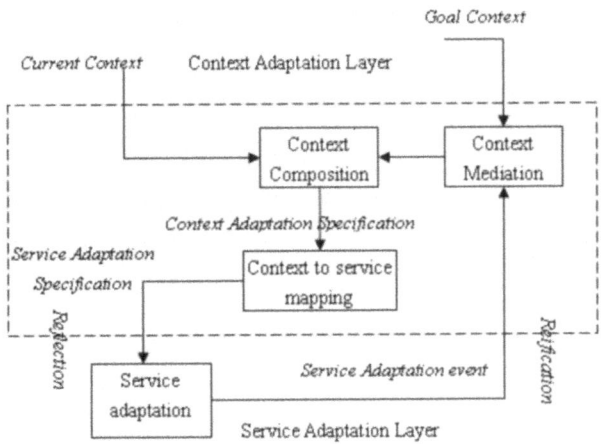

Fig. 3. The reflective middleware layer model in more detail

to the service adaptation specification, the representation of the actual service adaptation. Service adaptation is treated as the base system to be reflected on. The service adaptation specification defines policies and parameters for service adaptation.

While a service is being adapted, internal events may occur because of unexpected inconsistencies arising from external context changes. These events will be reported to the context mediator in the AML, which mediates between the context changes and the context composition leading to more realistic goal contexts. Mediation can occur in two ways: the goal contexts can be mediated automatically if they are represented semantically, e.g., short-time travel can sometimes be viewed as short-distance travel if the travel speed is not important. If the mediation cannot be done automatically, user intervention is needed to revise the goal contexts. Any changes taking place in the context composition will lead to a change in the service adaptation specification, which therefore reflects the change back to the service itself.

3 Application Trial and Discussion

A map demonstrator [9] has been implemented by using Geotools [10] where our system has been deployed to acquire some results from test cases shown in figure 4. When re-route is required for a traffic block at the point of Queen Mary, the case on the left shows the pre-planed route R1, and the other one shows the adapted route R2. R2 takes a longer distance and less scenic part of area, but will be much quicker to get to the destination in time. In R2 planning, the user goal contexts have been reconfigured, because the routing service first tried to get a route across more park areas which could maximize the aggregated benefits by the old weighting, but turned out to exceed the time limit; therefore the reflection is called to amend the weighting.

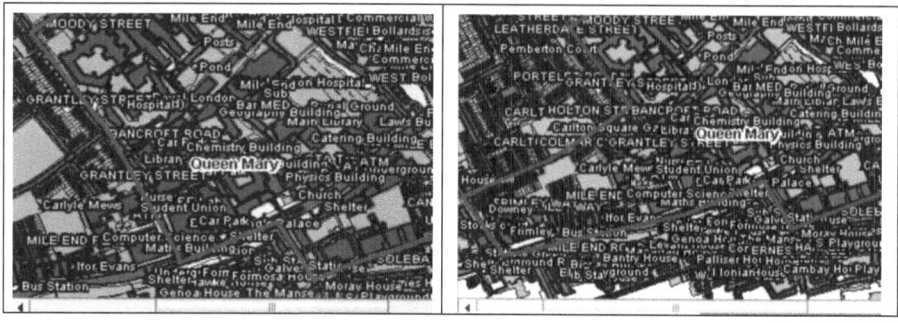

Fig. 4. A pre-planed static routing (left) and a reflective goal-based adaptive routing (right)

4 Conclusion

We investigated and applied a reflective goal-based system for context-aware adaptation in a spatial routing application. The use of the goal context model and the multilateral adaptation framework can resolve the design issues as proposed. The system enables dynamic goal context configuration and hence service adaptation. In the future, we need to develop more use cases and to consider the overhead in deploying the system.

References

1. Fawcett, J., Robinson, P.: Adaptive routing for road traffic. IEEE Comput. Graph. Appl. 20(3), 46–53 (2000)
2. Rogers, S., Fiechter, C.N., Langley, P.: An adaptive interactive agent for route advice. In: The third annual conference on Autonomous Agents, Seattle, Washington, United States (1999)
3. Hochmair, H.: Towards a Classification of Route Selection Criteria for Route Planning Tools. In: Developments in Spatial Data Handling 11th International Symposium on Spatial Data Handling (2005)
4. Lee, J., Park, S., et al.: ACE-INPUTS: A Cost-Effective Intelligent Public Transportation System. IEICE Trans. Inf. & Syst. E90–D(8), 1251–1261 (2007)
5. Dey, A.K., Abowd, G.D.: Towards a better understanding of context and context-awareness. In: Proceedings of the Workshop on the What, Who, Where, When and How of Context-Awareness. ACM Press, New York (2000)
6. Meng, D., Poslad, S.: A reflective context-aware system for spatial routing applications. In: MPAC 2008, pp. 54–59 (2008)
7. Poslad, S.: Ubiquitous Computing: Smart Devices, Environments and Interactions. Wiley, Chichester (2009)
8. Maes, P.: Concepts and Experiments in computational Reflection. In: OOPSLA 1987 Proceedings, pp. 147–155 (1987)
9. Liang, Z., Poslad, S., Meng, D.: Adaptive Sharable Personalized Spatial-Aware Map Services for Mobile Users. In: The GI-Days 2008 Conference, Munster, German, pp. 267–273 (2008)
10. Geotools The open source Java GIS toolkit, http://geotools.codehaus.org/

Pervasive Adaptation in Car Crowds

Alois Ferscha and Andreas Riener

Department for Pervasive Computing,
Johannes Kepler University, A-4040 Linz, Austria
Tel.: +43(0)732/2468/8555, Fax: +43(0)732/2468/8425
{ferscha,riener}@pervasive.jku.at

Abstract. Advances in the miniaturization and embedding of electronics for microcomputing, communication and sensor/actuator systems, have fertilized the pervasion of technology into literally everything. Pervasive computing technology is particularly flourishing in the automotive domain, exceling the "smart car", embodying intelligent control mechanics, intelligent driver assistance, safety and comfort systems, navigation, tolling, fleet management and car-to-car interaction systems, as one of the outstanding success stories of pervasive computing. This paper raises the issue of the socio-technical phenomena emerging from the reciprocal interrelationship between drivers and smart cars, particularly in car crowds. A driver-vehicle co-model (DVC-model) is proposed, expressing the complex interactions between the human driver and the in-car and on-car technologies. Both explicit (steering, shifting, overtaking), as well as implicit (body posture, respiration) interactions are considered, and related to the drivers vital state (attentive, fatigue, distracted, aggressive). DVC-models are considered as building blocks in large scale simulation experiments, aiming to analyze and understand adaptation phenomena rooted in the feed-back loops among individual driver behavior and car crowds.

Keywords: Pervasive Adaptation, Socio-technical Systems, Smart Cars, Car Crowds, Driver-Car Interaction, Vital Context.

1 Car Crowds as Socio-Technical Systems

The term and notion of *socio-technical systems* emerged from the context of labor studies, conducted around the early sixties [1]. Labor studies, generally concerned with the adaptation of humans to organizational and technical frameworks of work or production, analyzed the impact of the *"human factor in industrial relations"*, like e.g. in manufacturing systems proposed by Henry Ford or Frederick Winslow Taylor, and attempted to understand the interrelationship among humans and machines from both the technical (*"efficiency"*) as well as the social (*"humanity"*) conditions of work. A considerable body of research on socio-technical systems emerged as *"organizational development"* (R. Beckhard, MIT Sloan School of Management), addressing the principles and techniques of harmonizing complex organizational work design (*"humanization of work"*) with

C. Hesselman and C. Giannelli (Eds.): Mobilware 2009 Workshops, LNICST 12, pp. 111–117, 2009.

the optimization of productivity. All this research roots on the recognition of the *interaction* between people and technology in the workplace.

More modern socio-technical systems research has looked into the principles and properties of systems considered complex at the confluence of society and technology, particularly the principles and properties that make a system -constituted of many elements that interact to produce "global" behaviour- exhibit a "global" behavior that cannot (easily) be explained in terms of the interactions among the individual elements. More generally, and on a very abstract level, complexity science [2] attempts to better understand systems in which aggregate, system-level behaviour arises from the interactions between component parts in a way that is not straightforward. Such *complex systems* are described as *"... a dynamic network of many agents (which may represent cells, species, individuals, firms, nations) acting in parallel, constantly acting and reacting to what the other agents are doing* where the control *tends to be highly dispersed and decentralized,* and if there is to be *any coherent behavior in the system, it has to arise from competition and cooperation* among the agents, so that the *overall behavior of the system is the result of a huge number of decisions made every moment by many individual agents.".* Conclusively, a complex *adaptive* systems is one in which either (*i*) the number of elements (or parts of the system) and the relations among them are non-trivial (or non-linear), and/or (*ii*) the system has memory or feedback, and/or (*iii*) the relations between the system and its environment are non-trivial (or non-linear), and/or (*iv*) the system can be influenced by, or can adapt itself to a situation or the environment, and/or (*v*) the system is highly sensitive to initial conditions.

In order to study emergent behaviour and phenomena of self organization in complex road traffic scenarios, and following the lines of a socio-technical analysis of the phenomena emerging in traffic, we can assume car crowds as complex adaptive systems (CASs) due to the following observations:

- The interaction among cars in a traffic scenario is seemingly random (since each and every car is following a "local" navigation goal, and the co-incidence of cars happening to come across each other on a certain road is unpredictable), while at the same time seemingly correlated (consider rush-hours, traffic jams or slack periods). The relations among arbitrary two cars in a car crowd is "non-trivial".

- The individual car behavior, as a consequence of the driver behavior, is impacted by memory and feedback. On one hand, routes (and jam escapes) that have been well learned will be repeated until there is an ultimate need to change them (memory). On the other hand, aggressive behaviour exhibited by drivers observably leads to arousal of other drivers, which again can cause aggressiveness (feedback).

- Considering time-of-day, day-of-week, weather conditions, road works or the such as the context [3] (or the "environment") of a car crowd "system", then the effect of such conditions is unpredictable ("non-trivial"). Crowd behaviour in road traffic situations does not change gradually, but changes abruptly after reaching a certain (unpredictable) "critical mass".

– Taking traffic lights as the means with which traffic can be "influenced", we observe "local adaptation" of individual cars/drives (like line-up allegiance phenomena), which at the same time, by completely disregarding surrounding or remote traffic situations, causes "global distortion" (local traffic jams).

With this work we attempt to lay ground for a complex adaptive systems analysis of car crowds, employing simulation based models. We start with models expressing the complex interactions among drivers and vehicles, so called *driver-vehicle co-models* (DVC-models), and arrange them as building blocks in large scale simulation experiments. The ultimate aim of such CAS simulation experiments is to analyze and understand adaptation phenomena rooted in the complex interactions and feed-back loops among individual driver behavior and car crowds in large traffic scenarios (10^5-10^7 entities).

1.1   Driver-Vehicle Co-Models

Modeling the interactions among a driver and the vehicle has to address two major aspects of complexity. First, on the driver side, it has to reflect the complex cognitive task of controlling the vehicle which is built up by four sub-processes (*i*) perception, (*ii*) analysis, (*iii*) decision, and (*iv*) expression (or simply as "chain of sensory perception") [4]. Second, on the vehicle side, data has to be gathered coming from sensors embedded into the car (implicit input), or from the respective controls (steering wheel, pedals, navigation panel, etc. explicit input). A particular difficulty when modeling the driver vehicle interaction loop is the orders of magnitude time discrepancies in the reaction of the driver and the the vehicle. Sensor based data recording, instrumentation, and processing on the vehicle input side, as well as actuator control on the vehicle output side by far excels the human perception-analysis-decision-expression process [5]. Driver Assistance Systems (DAS) have emerged, aiming to improve (power steering) or compensate (ABS breaking) driver performance, but potentially elevate cognitive load at the same time. In addition, on-board entertainment systems can lead to an overload of the visual or auditory channels of perception, again having negative impact onto reaction time. Last, but not at least do vital parameters like fatigue, stress, attention, etc. crucially affect driver performance. All these aspects are essential aspects to be represented in a DVC-model.

As a first approach towards a DVC-model we have therefore focused on the vital state of a driver, and the technological means to continuously collect data so as to be able to compute what we call the *driver vital context* (see Figure 1).

Here the vital context of a driver is an aggregate of information coming from physical sensors, capturing the physiological (heart rate, heart rate variability, skin conductance, body temperature, respiration frequency, etc.) and cognitive (workload, stress, fatigue, etc.) attributes relevant for the analysis of the driver-vehicle interaction loop. In first experiments we have focused on the parameters extractable from an electrocardiogram (ECG)) which are in particular the (*i*) heart rate (RR), (*ii*) heart rate variability (HRV), (*iii*) "autochronic image" (AI) [6] and (*iv*) standard deviation of normal RR intervals (SDNN) as indicated in

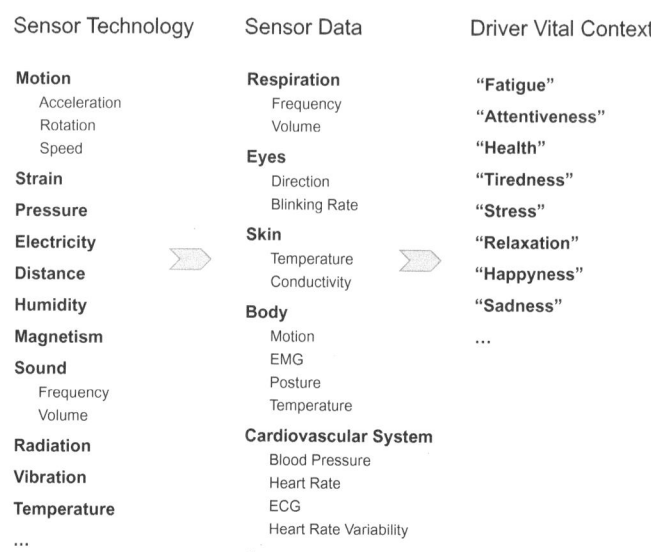

Fig. 1. Computing the driver vital context from sensor data

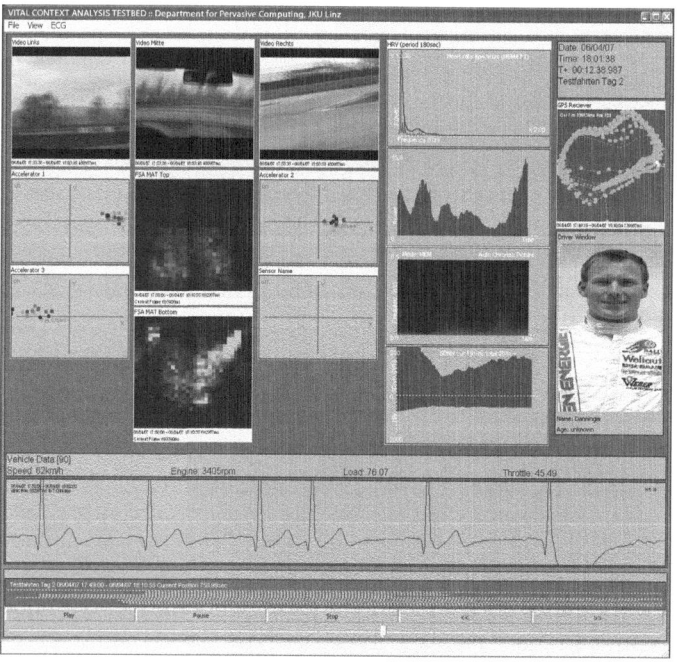

Fig. 2. The vital context analysis testbed allows to analyze correlations between vehicle-specific data and a driver's vital context

the fourth column of Fig. 2 (from top to bottom) and have used them in our "vital context analysis testbed" [7] [8] framework to answer questions regarding the interrelationship between a driver's vital context (or "mood") and specific driving situations. A specific source of "mood" indication is the autochronic image (Fig. 2, column 4, third from top), a single feature representing the synopsis of (*i*) mood state, (*ii*) cognitive/mental workload, and (*iii*) activities of the autonomic nervous systems [9]. Much like the heart rate variability (HRV), the AI is frequently also considered as an indicator for the "emotional" state in chronobiology [10].

1.2   A Collective Driver-Vehicle Co-Model

The socio-technical issues we are interested in concerns the feed-back loop originating at the vital (or "emotional") state of a driver, directly translating into his driving style. Perceiving the driving style of other drivers, in turn, influences the emotional state and hence driving style of hte observers. These transitional, yet collective driver state and driving style changes raise global car crowd phenomena like traffic jams, collective aggressiveness, lane blocking, etc.

Often driving style is communicated to nearby cars only, and implicitly [11] as it is being observed by other drivers. A *collective DVC-model*, therefore, needs to reflect this propagation of information within constrained local boundaries appropriately. Diffusion of driver state information to "neighboring" cars, or within field-of-view ranges appears appropriate to address global car crowd phenomena with large scale simulation experiments. Figure 3 sketches the architecture of a complex car crowd model with either centralized or decentralized information management logic. The vehicle-drive interaction loop is extended in the sense that the sensors recognized driver state is propagated into a collective

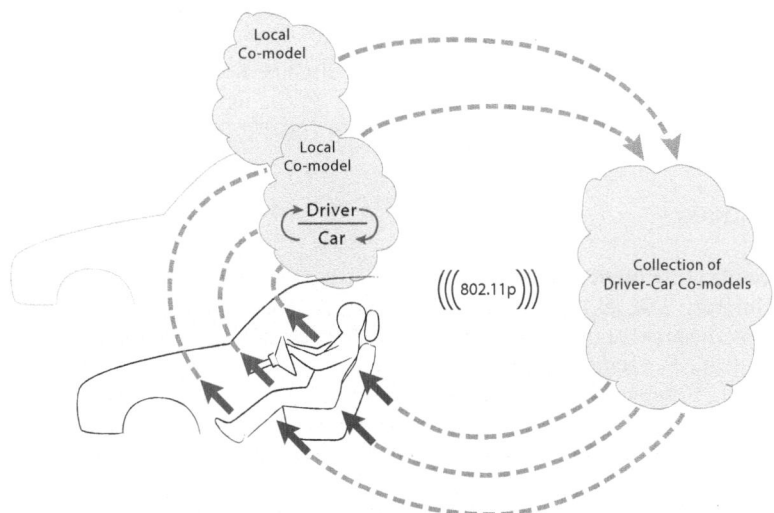

Fig. 3. The collective Driver-Vehicle Co-Model is built from local DVC-models

DVC-model, which in turn generates a flow of control information back to the individual driver.

2   Conclusions and Further Work

Pervasive computing technologies have revolutionized the car driving experience, with significant advances in car steering, breaking and accelerating, etc., navigating, communicating, safety, driving comfort and even entertainment. While most of these advances concern a single driver or a single car, a whole lot of potentials reside in the exploitation of technologies that let spontaneous groups of cars appear as a cooperative crowd (car-to-car communication, car-to-infrastructure communication, remote control, fleet management, etc.). Among the many scenarios that could gain from such technologies are congestion avoidance, traffic shaping, environment protection, energy preservation, power saving, etc.

Car crowds represent, however, cases of complex adaptive systems, in which aggregate, system-level (or global) behaviour arises from frequent and complex (local) interactions between component parts in a way that is "non-trivial". In order to study phenomena emerging from such complex system behavior we attempt for models suitable for simulation based analysis. Particularly for car crowds we propose a driver-vehicle co-model abstracting "local" interactions, and a collective driver-vehicle co-model abstracting "global" behavior. The DVC-model builds upon the driver state and in-car interactions, whereas the CDVC-model expresses cross-car and car-to-infrastructure interactions. Large scale simulation experiments [12] [13] involving these models will gain insight in the mechanism of vehicular pervasive adaptation.

Acknowledgements

This work is supported under the FP7 ICT Future Enabling Technologies programme of the European Commission under grant agreement No 231288 (SOCIONICAL) and grant agreement No 225938 (OPPORTUNITY).

References

1. Ropohl, G.: Philosophy of Socio-technical Systems. Society for Philosophy and Technology. Digital Library and Archives 4(3) (Spring 1999),
   http://scholar.lib.vt.edu/ejournals/SPT/v4_n3html/ROPOHL.html
2. Castellani, B., Hafferty, F.W.: Sociology and Complexity Science. Series: Understanding Complex Systems. Springer, Heidelberg (2009)
3. Ferscha, A., Vogl, S., Beer, W.: Context sensing, aggregation, representation and exploitation in wireless networks. Parallel and Distributed Computing 6(2), 77–81 (2005)
4. Riener, A.: Sensor-Actuator Supported Implicit Interaction in Driver Assistance Systems. Phd thesis, Department for Pervasive Computing, Johannes Kepler University Linz, Austria (2009)

5. Kopetz, H.: Real-Time Systems. Kluwer Academic Publishers, Dordrecht (1997)
6. Moser, M., Schaumberger, K., Fruehwirth, M., Penter, R.: Chronomedizin und die neue Bedeutung der Zeit. Promed Komplementr, 8–18 (October 2005)
7. Ferscha, A., et al.: Context Framework for Mobile Devices (CON): Vital Context. Industrial Cooperation Siemens AG Munich. Institute for Pervasive Computing, JKU Linz (Presentation Documents), September 26 (2007)
8. Riener, A., Ferscha, A., Matscheko, M.: Intelligent vehicle handling: Steering and body postures while cornering. In: Brinkschulte, U., Ungerer, T., Hochberger, C., Spallek, R.G. (eds.) ARCS 2008. LNCS, vol. 4934, pp. 68–81. Springer, Heidelberg (2008)
9. Moser, M., Lehofer, M., Sedminek, A., Lux, M., Zapotoczky, H.G., Kenner, T., Noordergraaf, A.: Heart Rate Variability as a Prognostic Tool in Cardiology A Contribution to the Problem From a Theoretical Point of View. Circulation 90(2), 1078–1082 (1994)
10. Moser, M., Fruhwirth, M., Kenner, T.: The Symphony of Life [Chronobiological Investigations]. IEEE Engineering in Medicine and Biology Magazine 27(1) (January-Feburary 2008)
11. Ferscha, A.: Implicit Interaction. In: The Universal Access Handbook. Lawrence Erlbaum Associates, Inc., Mahwah (2009)
12. Ferscha, A.: Parallel and distributed simulation of discrete event systems. In: Zomaya, A.Y. (ed.) Parallel and Distributed Computing Handbook, pp. 1003–1041. McGraw-Hill, New York (1996)
13. Ferscha, A., Johnson, J., Turner, S.J.: Distributed simulation performance data mining. Future Generation Comp. Syst. 18(1), 157–174 (2001)

Developing User-Centric Applications with H-Omega

Clement Escoffier[1], Jonathan Bardin[2], Johann Bourcier[3], and Philippe Lalanda[2]

[1] akquinet AG, Bülowstraße 66, 10783 Berlin, Germany
`clement.escoffier@akquinet.de`
[2] Grenoble University, BP 53, 38041, Grenoble Cedex 9, France
`{jonathan.bardin,philippe.lalanda}@imag.fr`
[3] Imperial College London, 180 Queens Gate, London SW7 2BZ, UK
`jbourcier@doc.ic.ac.uk`

Abstract. The recent proliferation of ever smaller and smarter electronic devices, combined with the introduction of wireless communication and mobile software technologies enables the construction of a large variety of pervasive applications. The inherent complexity of such applications along with their non-expert clientele raises the necessity of building middleware solutions. This paper proposes to use the H-Omega application server to build user-centric applications. H-Omega relies on a service oriented component platform, iPOJO, and provides useful technical services for pervasive applications. This paper presents how to easily develop user-centric applications on the top of the H-Omega platform.

Keywords: pervasive application, application server, web portal, iPOJO.

1 Introduction

Our present living environments are being increasingly populated with ever smarter and smaller electronic devices. The introduction of such communicating devices has already changed the way we interact with social and physical environment [1]. However, this trend grows up, and it seems to be a mere beginning. Devices continue to feature progressively more and more sophisticated functionalities, and become to interact each other for providing new, higher-level services.

Most of the research efforts already done were focusing on providing hardware that can actually enable such interactions. Plenty of devices providing this kind of features are already commercialized. But very few interesting applications take advantages of this new infrastructure. This is mainly due to the complexity of building software that can actually benefit from this underlying hardware. Indeed, usual software engineering techniques and tools are not suitable. Several software engineering challenges remain to be solved before fulfilling the vision of a pervasive world. Among these challenges, we can notice the high degree of dynamics, distribution, and heterogeneity of the devices involved and also major security and privacy concerns.

We have investigated a global architecture for designing pervasive applications especially related to the home environment [2, 3]. The main result of our previous works was the design of a residential application server providing a suitable runtime

C. Hesselman and C. Giannelli (Eds.): Mobilware 2009 Workshops, LNICST 12, pp. 118–123, 2009.

for pervasive applications. This runtime is based on a service-oriented component platform, a set of common services and a mechanism reifying all available devices as services. This platform considerably simplifies the development of pervasive application as stated in [3, 4]. This paper describes how this infrastructure can be used to develop user-centric applications.

The rest of the paper is organized as follows. First, we present the background of this work, including pervasive computing challenges, and the service-oriented computing paradigm. This is followed by a description of the H-Omega application server and how it resolves the common pervasive application development issues. Examples of user-centric applications will also be presented. Finally this paper will conclude by pointing out major contributions, and ongoing work.

2   Requirements and Background

2.1   Requirements for Successful Pervasive Computing

The success of any pervasive system highly depends on several key elements, including provided functionalities, Quality of Service (QoS) and affordability. In short, pervasive applications must offer services that are useful, or somehow interesting to the user. In addition, provided functionality must be associated with domain-specific QoS guarantees, such as performance, dependability and usability. System performance allows users to experience natural, *real-time* interactions with the pervasive environment. Dependability ensures service reliability and security and implies that the same behavior is experienced every time the system is run in similar execution scenarios. Application usability implies ease of service exploitation, with no expert knowledge required and with minimal maintenance effort necessary for modifying and evolving the system. Finally, the overall utility of provided services must overcome the effort required to acquire and maintain the corresponding pervasive systems. Affordable pervasive solutions imply that clients are willing to invest the required resources in exchange for offered functionalities, both in the short term (e.g. acquisition and installation) and over long durations (e.g. maintenance).

Pervasive computing systems generally consist of various electronic devices and software entities capable of communicating with one another. Different types of software-equipped appliances may be available for a variety of purposes, such as interacting with the real environment, providing control and display services, or exposing data and application interfaces to other devices and applications. The main challenge of the pervasive computing domain is to provide coherent pervasive environments, offering useful applications and services, based on an entanglement of heterogeneous, distributed and dynamic devices and software services, communicating via various technologies and protocols. In this context, several characteristics specific to pervasive equipments make this domain appealing from a business perspective, while raising difficult problems for system development and maintenance. Such device properties include:

- Distribution. Devices are typically scattered across the physical environment and accessible via diverse protocols, generally over a wireless communication support.

- Heterogeneity. A vast range of appliances, software technologies and communication protocols are available in the pervasive computing domain. A consensus on uniform and compatible implementations is not presently foreseen.

- Limited resources. Resource availability is generally scarce on the physical execution platforms employed in pervasive systems. In the pervasive home context, software applications typically run on a small gateway, with little memory space and low processing capabilities.

- Dynamism. Device availability is by far most volatile in pervasive systems with respect to other computing system types. This is due to several facts, including: i) users may freely and frequently change their locations and hence the locations of equipments they carry; ii) users may voluntarily activate and deactivate devices, or devices may unexpectedly run out of battery. This directly impacts on the availability of services running on these devices; iii) users and providers may periodically update deployed software services.

In addition to hardware and software dynamism, pervasive systems are constantly confronted with changes in their execution contexts. This may include modifications in the user's current behavior, social circumstance, location, mood, or general routine, as well as changes in other software applications' availability and behaviors.

Nowadays, there is no frameworks dealing with all those requirements and providing facilities to design, develop, deploy and execute pervasive applications. Such framework must provide a way to create dynamic and secure (safety, privacy...) applications dealing with heterogeneous devices and services.

2.2 Service-Oriented Computing

Service-Oriented Computing (SOC) [5, 6] is a new trend in software engineering that uses services as basic elements for building applications. In this computing paradigm, a service represents a computational entity described by a specification. This specification contains both functional (service interface) and non-functional (QoS) parts, without referring to the service implementation. Consequently, services can be supplied by multiple service providers and feature various implementations. At runtime, available services are made accessible by registering into service brokers where service consumers can dynamically discover them. A service consumer is then able to invoke the service by only relying on the service specification.

iPOJO [7, 8] is a service oriented component runtime that aims to simplify the development of applications on top of OSGi [9] platforms. iPOJO allows the straightforward development of application logic based on Plain Old Java Objects (POJO). iPOJO subsequently injects non-functional facilities into the application components, as necessary. Such facilities cover service provisioning, service dependency and lifecycle management. So components are bound by following the service-oriented interaction pattern. So, they can be developed and evolve separately. In addition to providing a reusable set of non-functional capabilities, iPOJO is seamlessly extensible to include new middleware functionalities. The iPOJO framework merges the advantages of component and service oriented paradigms. Specifically, iPOJO application functionalities are implemented following the component orientation paradigm. Each component is fully encapsulated, self-sufficient and provides server and client interfaces exposing its functionalities and dependencies, respectively.

3 H-Omega: An Application Server for Pervasive Applications

In our previous works, we developed an infrastructure to execute pervasive applications on residential gateways. The resulting infrastructure, named H-Omega, is an application server targeting specifically pervasive applications. This work follows a software engineering trend that aims to build flexible and modular application servers providing suitable runtime environments for applications such as JEE and Spring. The existing application servers do not suit to our particular application domain, which requires specific technical services, dynamism and flexibility. The main part of the H-Omega architecture is a residential gateway that could be seen as a home application server. Residential applications are deployed and run on this gateway. This gateway has the ability to discover and interact with dynamic devices inside the home and also with external services.

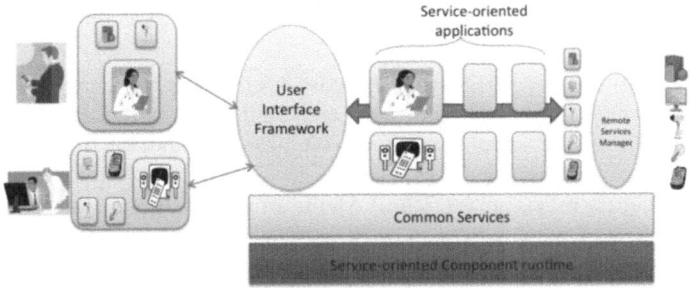

Fig. 1. H-Omega architecture

The figure 1 shows the internal design of the H-Omega residential server. This computing infrastructure is based on the service-oriented computing paradigm. Home applications are built using the iPOJO service oriented component model. All available devices are reified as services in the framework. The life cycle of these services follows the availability of the device they represent. These services take in charge the communication between applications and the real devices or web services. Application developer can then rely on these services to build their own application without taking care of all the tricky problems of device distribution, heterogeneity and dynamism.

The H-Omega server also provides common services in order to further simplify the development of residential application. These common services are required functionalities across applications. These provided facilities currently include event communication, scheduling of tasks, data persistence and remote administration facilities. To support user-centric applications, H-Omega also provides an infrastructure to monitor the context. This context contains information related to inhabitants (location, medical data), as well as physical metrics (temperature, luminosity…). The context itself is an aggregation of context sources that can be sensors, or applications. Applications can also be notified when the context changes and react to context changes. Applications interested in the context just specify the monitored properties thanks to a LDAP filter, and will be automatically notified when a change occurs changing the LDAP filter evaluation. Each application can also defines its own private context service containing only the domain-specific data (in the domain language). This private service creates a bridge between context sources and the global context service and the application context. Thanks to this context service, applications can directly collect "understandable" data and be notified of meaningful changes.

The H-Omega server[1] constitutes an open infrastructure in which service providers can freely deploy and withdraw applications, taking advantages of the available devices in the home and of the context. The H-Omega framework copes the requirements of pervasive environment and provides a way to design, develop, execute and manage pervasive applications. The H-Omega server was developed and used in the ANSO ITEA project. Moreover, it is still used by France Telecom and Schneider Electric. Akquinet A.G. also investigates using the H-Omega server for mobile applications.

4   The Follow Media Application

To illustrate how to use H-Omega to create user-centric application, this section describes a media-on-demand application selecting the media render according to the user location. Despite this application focus on media, it is not the only purpose of the H-Omega gateway.

This application relies on the UPnP Media profile [10] and the context service provided by the H-Omega server to track user location. Medias are stored in UPnP media servers (Figure 2), or in any media repository reified as a media server service in the gateway (for example a video-on-demand Internet service). As soon as a new media server service is available in the house, contained media will be added to the application. The user can list available music and movies. When the user begins to listen or watch a media, the application selects the media renderer that is in the same room as the user, and plays the media to this renderer. If the user moves to another room, the media is stopped. As soon as the user is back in a room with another UPnP media renderer, the application continues to play the media on this renderer.

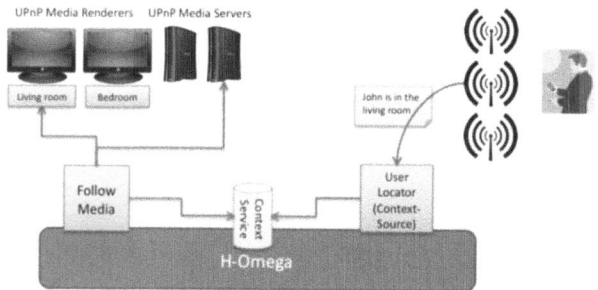

Fig. 2. Follow Media Application

---

[1] H-Omega is an open-source project available at http://ligforge.imag.fr/projects/homega/

Thanks to the H-Omega application server, the application knows at anytime where is located the user and which media server and renderer are available. One of the main advantages of this application is that it can interact with heterogeneous repositories. UPnP Media service are obviously supported but the application maps them as "raw" media servers. So, any other repository exposed as a "raw" media server can be used by the application.

5 Conclusion

Despite the democratization of smart objects, developing pervasive applications is far from easy. The H-Omega application server provides a suitable runtime to develop pervasive application for home context applications. Moreover, H-Omega provides abilities for applications to track context changes. This paper has demonstrated how H-Omega can be used to create user-centric applications. Two examples were presented.

We are currently investigating several perspectives. First, we are building an autonomic toolkit on the top of H-Omega to create applications with a higher-degree of autonomy. Moreover, we are considering the generation of domain-specific IDE to help the design, development, and management of home-context applications.

References

1. Weiser, M.: The computer for the 21st century. Scientific American 265(3) (1991)
2. Bourcier, J., Escoffier, C., Lalanda, P.: Implementing home-control applications on service platform. In: 4th IEEE Consumer Communications and Networking Conference (CCNC 2007), Las Vegas (January 2007)
3. Escoffier, C., Bourcier, J., Lalanda, P.: Toward an Application Server for Home Applications. In: 5th IEEE Consumer Communications and Networking Conference (CCNC 2008), Las Vegas (2008)
4. Bottaro, A., Bourcier, J., Escoffier, C., Lalanda, P.: Context-Aware Service Composition in a Home Control Gateway. In: IEEE International Conference on Pervasive Services (2007)
5. Papazoglou, M.P., Georgakopoulos, D.: Service-oriented computing. Commun. ACM 46(10), 24–28 (2003)
6. Huhns, M.N., Singh, M.P.: Service-Oriented Computing: Key Concepts and Principles. IEEE Internet Computing 9, 75–81 (2005)
7. Escoffier, C., Hall, R.S., Lalanda, P.: iPOJO An extensible service-oriented component framework. In: IEEE International Conference on Service Computing. Salt Lake City (2007)
8. Escoffier, C., Hall, R.S.: Dynamically adaptable applications with iPOJO service components. In: Lumpe, M., Vanderperren, W. (eds.) SC 2007. LNCS, vol. 4829, pp. 113–128. Springer, Heidelberg (2007)
9. OSGi Alliance. OSGi Service Platform Core Specification Release 4 (August 2005), http://www.osgi.org
10. The UPnP Forum, http://www.upnp.org

Utilization Possibilities of Area Definition in User Space for User-Centric Pervasive-Adaptive Systems

Ondrej Krejcar

VSB Technical University of Ostrava, Centre for Applied Cybernetics, Department of measurement and control, 17. Listopadu 15, 70833 Ostrava Poruba, Czech Republic
Ondrej.Krejcar@remoteworld.net

Abstract. The ability to let a mobile device determine its location in an indoor environment supports the creation of a new range of mobile information system applications. The goal of my project is to complement the data networking capabilities of RF wireless LANs with accurate user location and tracking capabilities for user needed data prebuffering. I created a location based system enhancement for locating and tracking users of indoor information system. User position is used for data prebuffering and pushing information from a server to his mobile client. All server data is saved as artifacts (together) with its indoor position information. The area definition for artifacts selecting is described for current and predicted user position along with valuating options for artifacts ranging. Future trends are also discussed.

Keywords: Prebuffering, Localization, PDPT Framework, Wi-Fi, Mobile Device, Area Definition.

1 Introduction

The usage of various mobile wireless technologies and mobile embedded devices has been increasing dramatically every year and would be growing in the following years. This will lead to the rise of new application domains in network-connected PDAs (Personal Digital Assistants) that provide more or less the same functionality as their desktop application equivalents. Context is relevant to the mobile user, because in a mobile environment the context is often very dynamic and the user interacts differently with the applications on his mobile device in different context [1].

My recent research of context-aware computing has been restricted to location-aware computing for mobile applications using a WiFi network (LBS Location Based Services). The information about basic concept and technologies of user localization such as LBS, Searching for WiFi AP) can be found in my article [2]. On localization basis, I created a special framework called PDPT (Predictive Data Push Technology) which can improve a usage of large data artifacts of mobile devices [3]. I used continual user position information to determine a predictive user position. The data artifacts linked to user predicted position are prebuffered to user mobile device. When user arrives to position which was correctly determined by PDPT Core, the data artifacts are in local memory of PDA. The time to display the artifacts from local memory is much shorter than in case of remotely requested artifact.

C. Hesselman and C. Giannelli (Eds.): Mobilware 2009 Workshops, LNICST 12, pp. 124–130, 2009.

The prebuffering techniques may not be an only one application method for user position knowledge. I would like to describe a predictive position determination as well as area definition background in next chapter to give a reader more information about these themes to discuss new utilization possibilities in third chapter.

2 The PDPT Framework and PDPT Core

The general principle of my simple localization states that if a WiFi-enabled mobile device is close to such a stationary device – Access Point (AP) it may "ask" the provider's location position by setting up a WiFi connection. If position of the AP is known, the position of mobile device is within a range of this location provider. This range depends on type of WiFi AP. The Cisco APs are used in my test environment at Campus of Technical University of Ostrava. I performed measurements on these APs to get signal strength (SS) characteristics and a combination of them called "super ideal characteristic". More details can be found in chapter 2.3 [5]. The computed equation for Super-Ideal characteristic is taken as basic equation for PDPT Core to compute the real distance from WiFi SS. From this super ideal characteristic it is also evident the signal strength is present only to 30 meters of distance from base station. This small range is caused by using of Cisco APs. These APs has only 2 dB WiFi omnidirectional antenna. Granularity of location can be improved by triangulation of two or more visible WiFi APs. The PDA client will support the application in automatically retrieving location information from nearby location providers, and in interacting with the server. Naturally, this principle can be applied to other wireless technologies like Bluetooth, GSM or WiMAX. To let a mobile device determine its own position is needed to have a WiFi adapter still powered on. This fact provides a small limitation of use of mobile devices. The complex test with several types of battery is described in my article [4] in chapter (3). The test results with a possibly to use a PDA with turned on WiFi adapter for a period of about 5 hours.

2.1 Predictive Data Push Technology

PDPT framework is based on a model of location-aware enhancement, which I have used in created system. This technique is useful in framework to increase the real dataflow from wireless access point (server side) to PDA (client side). Primary dataflow is enlarged by data prebuffering. PDPT pushes the data from SQL database to clients PDA to be helpful when user comes at final location which was expected by PDPT Core. The benefit of PDPT consists in time delay reducing needed to display desired artifacts requested by a user from PDA. This delay may vary from a few seconds to number of minutes. Theoretical background and tests were needed to determine an average artifact size for which the PDPT technique is useful. First of all the maximum response time of an application (PDPT Client) for user was needed to be specified. Nielsen [6] specified the maximum response time for an application to 10 seconds [7]. I used this time period (10 second) to calculate the maximum possible data size of a file transferred from server to client (during this period). If transfers speed wary from 80 to 160 kB/s the result file size wary from 800 to 1600 kB. More details can be found in chapter 2.5 [5]. The next step was an average artifact size definition. I use a network architecture building plan as sample database, which contained 100 files of average size of 470 kB. The client application can download during the 10 second period from 2 to 3 artifacts. The final result of my real tests and consequential calculations is definition of artifact size to average value of 500 kB. The buffer size may differ from 50 to 100 MB in case of 100 to 200 artifacts.

2.2 The PDPT Framework Design

The PDPT framework design is based on the most commonly used server-client architecture. The PDPT framework server is created as a web service to act as a bridge between MS SQL Server (other database server eventually) and PDPT PDA Clients [Fig. 1].

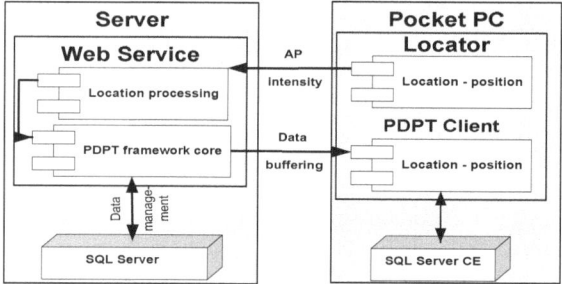

Fig. 1. PDPT architecture – UML design

Client PDA has location sensor component, which continuously sends the information about nearby AP's intensity to the PDPT Framework Core. This component processes the user's location information and it makes a decision to which part of MS SQL Server database needs to be replicated to client's SQL Server CE database. The PDPT Core decisions constitute the most important part of PDPT framework, because the kernel must continually compute the position of the user and track, and predict his future movement. After doing this prediction the appropriate data are prebuffered to client's database for the future possible requirements. This data represent artifacts list of PDA buffer imaginary image [Fig. 2].

2.3 PDPT Core – Static and Dynamic Area Definition

The PDPT buffering and predictive PDPT buffering principle is shown in [Fig. 2]. Firstly the client must activate the PDPT on PDPT Client. This client creates a list of artifacts (PDA buffer image), which are contained in his mobile SQL Server CE database. Server create own list of artifacts (imaginary image of PDA buffer) based on area definition for actual user position and compare it with real PDA buffer image.

The area can be defined as an object where the user position is in the object centre. I am using the cuboid as the object in present time for initial PDPT buffering. This cuboid has static area definition with a size of 10 x 10 x 3 (high) meters. The PDPT

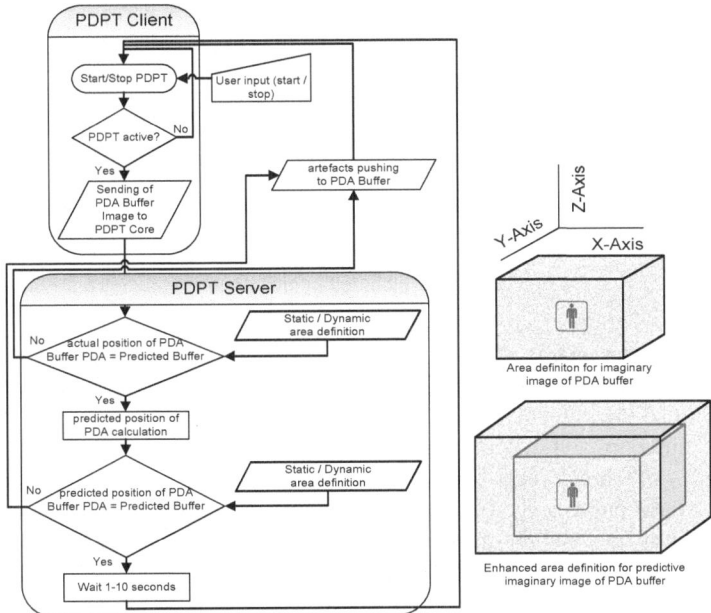

Fig. 2. Object diagram of PDPT prebuffering and predictive PDPT prebuffering. Right part shows the area definition for imaginary image of PDA buffer.

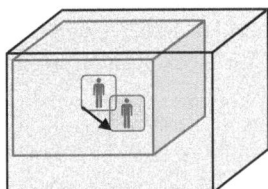

Fig. 3. Area enlargement to direction of predicted user position

Core will continue with comparing of both images. In case of some difference, the rest artifacts ale prebuffered to PDA buffer. When all artifacts for current user position are in PDA buffer, there is no difference between images. In such case the PDPT Core is going to make a predicted user position. On base of this new user position it makes a new predictive enlarged imaginary image of PDA buffer.

The size of this new cuboid is statically defined area of size 20 x 20 x 6 meters. The new cuboid has a center in direction of predicted user moving and includes a cuboid area for current user position [Fig. 3]. The PDPT Core compares the both new images (imaginary and real PDA buffer) and it will continue with buffering of artifacts until they are same. The process of selecting the artifacts for imaginary image of PDA buffer consists of finding and evaluating of artifacts inside the specified area in 3D environment [Fig. 4].

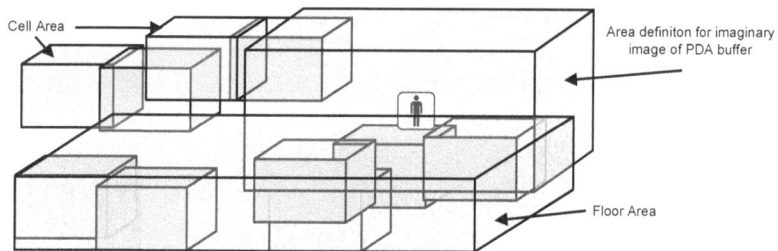

Fig. 4. Selecting artifacts which are included in area definition for current user position

1st, Corners numbering – Var. A: First possibility of evaluation is in corners numbering. The artifacts are stored in PDPT Server database along with position information of theirs corners. This information consists of six values: axis X, Y and Z, with the minimum and maximum values of artifact. First evaluation can be computed as a number of these corners inside the area.

2nd, Corners numbering – Var. B: Second one can be solve as intersection degree between the artifact and defined area. This way is more accurately, but it takes higher time consumption in most cases than previous one.

3rd, Priority counting: Third one possibility has a base in priority counting. Every artifact has own priority value which indicates a level of importance. The ranging of these levels can be made with all artifacts inside the area.

4th, Data types: Several types of data can be served as artifact. This property can be used as another way to valuating of artifacts.

First and third cases are used in PDPT Framework currently. The second case will be used in near future. The forth one cannot be currently used, because the only one type of data type is used for artifacts in sample PDPT database - the image type. Because of only two options are used and only a sample database exists, the static area definition is used now. In real case of usage is better to create an algorithm to dynamic area definition to adapt a system to user needs more flexible in real time.

2.4 The PDPT Client Application and Testing Results

The PDPT Client application realizes classical client to the server side and an extension by PDPT and Locator modules. Figure [Fig. 5] shows two screenshots from the mobile client. Figure [Fig. 5a] shows the classical view of the data artifact presentation from MS SQL CE database to user (in this case the image of Ethernet plan of the current area). The PDPT tab [Fig. 5b] presents a way to tune the settings of PDPT Framework. The middle section shows the logging info about the prebuffering process. The right side means measure the time of one artifact loading ("part time") and full time of prebuffering in millisecond resolution. More screens and details of PDPT Client can be found in chapters 2.7 and 2.8 [5].

I am focused on the real usage of the developed PDPT Framework on a wide range of wireless mobile devices and its main issue at increased data transfer. For testing purpose, five mobile devices were selected with different hardware and software capabilities. The high success rate found in the test data surpassed our expectations. This rate varies from 84 to 96 %. Please see the chapter 4 [5] for more info.

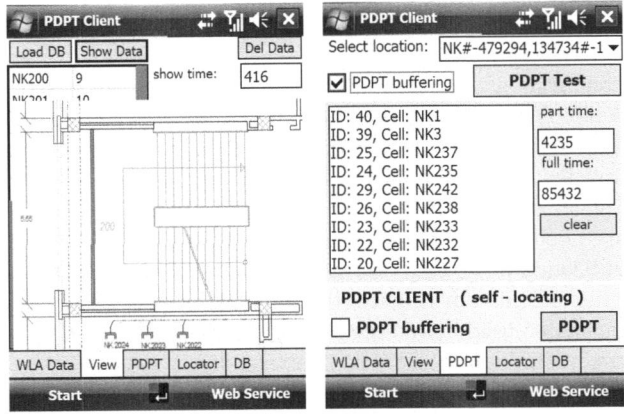

Fig. 5. PDPT Client – Left one figure 5a, Right one figure 5b

3   New Utilization Possibilities of Area Definition

The idea of user-centric pervasive-adaptive (UCPA) systems is in interaction between user and system throws his mobile device. Such interconnection can behold in the reaction on user's non declared requests. These requests include namely user current position, user future-predicted position, his movement and tracking (in case of my PDPT Framework). By the combination of these requests in conjunction with other sources of user's knowledge and behaviors, the sophisticated information system can be developed as UCPA system. The user comfort or theirs performance can be found namely in predicting the user needs along with the pre-reacting on them.

The larger and larger data amounts are transferred throw the internet network nowadays. The needs of techniques to reduce its amount or preloaded them before user needs is still growing up. PDPT Framework idea can be applied in a variety of wireless network systems now and in future naturally. More usability of PDPT grows from Area Definition as well as from evaluation of artifacts or other user's behavior sources. I can imagine the use in area of botanical or zoological gardens systems. The user needs to be located continually by WiFi or GPS. From current user position will the user track be computed online to allow make a prediction of user move. Data artifacts can be preloaded to user device memory for future requests. When user request info about his location in context of zoo or garden (switching on the device is only needed), the client application will respond with a map of near surroundings and it will start to play a documentary about animals or vegetation around the user. User can act with direct requests to selected kinds of these. These preferred kinds will be taken into account to evaluate future objects/artifacts and preloaded only the most important ones for user. The type of artifact is also evaluable as well as his size because user may not wont to look at too long or micro presentation. From several statistic info obtained from users tracks, the most frequented ways in gardens can be found. By the help of some mentioned info sources a very sophisticated dynamical area definition can be developed.

4 Conclusions

The PDPT prebuffering techniques can improve the using of medium or large artifacts on wireless mobile devices in information systems. If we can transform the real data from information system to artifacts along with their positions information, we can improve the transfer rate of used wireless connection and have a better response to users. The localization part of PDPT framework is currently used in another project of biotelemetrical system for home care named Guardian to make a patient's life safer [8]. Another utilization of PDPT consists in use of Area Definition. This idea can be utilized inside the information systems like botanical or zoological gardens where the GPS navigation can be used in outdoor. In combination with multi data type artifacts and dynamic area definition the new complex adaptable systems as well as the user-centric pervasive-adaptive systems can grown from.

Acknowledgment. This work was supported by the Ministry of Education of the Czech Republic under Project 1M0567.

References

1. Abowd, G., Dey, A., Brown, P., et al.: Towards a better understanding of context and context-awareness. In: Gellersen, H.-W. (ed.) HUC 1999. LNCS, vol. 1707, p. 304. Springer, Heidelberg (1999)
2. Krejcar, O.: User Localization for Intelligent Crisis Management. In: AIAI 2006, 3rd IFIP Conference on Artificial Intelligence Applications and Innovation, Athens, Greece, pp. 221–227 (2006)
3. Krejcar, O., Cernohorsky, J.: Database Prebuffering as a Way to Create a Mobile Control and Information System with Better Response Time. In: Bubak, M., van Albada, G.D., Dongarra, J., Sloot, P.M.A. (eds.) ICCS 2008, Part I. LNCS, vol. 5101, pp. 489–498. Springer, Heidelberg (2008)
4. Krejcar, O.: PDPT Framework - Building Information System with Wireless Connected Mobile Devices. In: ICINCO 2006, 3rd International Conference on Informatics in Control, Automation and Robotics, Setubal, Portugal, August 01-05, pp. 162–167 (2006)
5. Krejcar, O., Cernohorsky, J.: New Possibilities of Intelligent Crisis Management by Large Multimedia Artifacts Prebuffering. In: I.T. Revolutions 2008, Venice, Italy, December 17-19. LNICST. Springer, Heidelberg (2008)
6. Nielsen, J.: Usability Engineering. Morgan Kaufmann, San Francisco (1994)
7. Haklay, M., Zafiri, A.: Usability engineering for GIS: learning from a screenshot. The Cartographic Journal 45(2), 87–97 (2008)
8. Janckulik, D., Krejcar, O., Martinovic, J.: Personal Telemetric System – Guardian. In: Biodevices 2008, Insticc Setubal, Funchal, Portugal, pp. 170–173 (2008)

Architecture of a Personal Network Service Layer

Rieks Joosten[1], Frank den Hartog[1], and Franklin Selgert[2]

[1] TNO, Information- and Communication Technology
Brassersplein 2, P.O. Box 5050, 2600 GB Delft, The Netherlands
{Rieks.Joosten,Frank.denHartog}@tno.nl
[2] Koninklijke KPN N.V
Maanplein 55, P.O. Box 30000, 2500 GA Den Haag, The Netherlands
Franklin.Selgert@kpn.com

Abstract. We describe a basic service architecture that extends the currently dominant device-oriented approach of Personal Networks (PNs). It specifies functionality for runtime selection and execution of appropriate service components available in the PN, resulting in a highly dynamic, personalized, and context-aware provisioning of PN services to the user. The architectural model clearly connects the responsibilities of the various business roles with the individual properties (resources) of the PN Entities involved.

1 Introduction

Recently, a significant research effort has been made in the new concept of Personal Networks (PNs), most notably by the European FP6 projects MAGNET (BEYOND) and Ambient Networks [2]. Confusingly, both projects meant something slightly different with PNs. MAGNET defines a PN as "consisting of a core Personal Area Network (PAN) extended with clusters of remote devices which could be private, shared, or public and able to adapt to the quality of the network accessed" [3]. In a nutshell, this is about a user having remote access to his other devices. Ambient Networks, however, considered PNs more like defined by 3GPP, as "a PN consists of more than one Personal Network Element (PNE, i.e. either a single device or a group of devices) under the control of one user, providing access to the serving mobile network. The PNEs are managed in a way that the user perceives a continuous secure connection regardless of their relative locations [4]". This is more about redirecting sessions to the device that is most convenient to the user at a given time.

Both definitions have in common that they are oriented towards device access and device connectivity. One of the major results of the Dutch Freeband project PNP2008 is the understanding that, from a user perspective, the added value of PNs is more in accessing personal services and content, provided by either personal devices or other devices. Device access is then a *sine qua non*, but does not guarantee access to services offered by the device, nor does access to a personal service guarantee any control over the device which offers it.

This paper describes a basic architecture of a service layer extension for PNs as envisaged by MAGNET. As such we redefine PNs to include this service layer as

C. Hesselman and C. Giannelli (Eds.): Mobilware 2009 Workshops, LNICST 12, pp. 131–134, 2009.
© ICST Institute for Computer Sciences, Social-Informatics and Telecommunications Engineering 2009

follows: "a user-centric environment of potentially globally distributed resources, such as devices, services, content, network-nodes, etc." Devices, both within and outside PNs, host services and content that are dynamically configured to collaborate as individual applications whose functionality is enhanced by personal preferences and context information. The main challenge is to construct applications at runtime, from services that are technically available and the user is allowed to access from a business point of view.

2 Architecture

The architecture consists of PN Entities. PN Entities are instances of PN Entity Types, which are devices, network nodes, content, applications and other services. PN Entities are associated with the following PN business roles:

- Owners, who are responsible for issuing permissions and/or conditions under which the specific PN Entities they own may be used.
- Providers, who are responsible for the correct functioning of PN Entities at runtime including their operational management.
- Customers, who are responsible for the use of PN Entities within the context of a PN User Session, which may result in having to pay for that use.
- Users, who are individuals that within the context of a PN User Session actually use PN Entities.

The relations between these roles and PN Entities are graphically shown in the left part of Fig. 1. Notice that Customers, Providers and Owners may be organizations rather than individual persons, but Users may not. And only individuals can use or manage PN Entities, while organizations cannot. Even though organizations bear responsibilities, they always need a human individual that represents the organization to get something done.

Fig. 1 shows, on the right, the most relevant part of a functional architecture of a PN service layer. The central entity is the PN Administration Integrity Service (PNAIS). Its role is to preserve and provide guarantees with respect to the integrity of the administration data that resides in a potentially distributed database, the PN Provisioning Administration (PNPA). The integrity of the PNPA is defined as compliance to a set of constraints that have been specified explicitly for the PNPA, in terms of what defines the concepts (Device, PN Entity, User, etc.), and the relations between them. The PNAIS must enforce, in a runtime fashion, this set of constraints.

The User Agent & Authentication (UA) component manages the communication between individuals and their PNs, and ensures that PN Users only use services in their own PN. To achieve this, the UA must (a) interact with the user, (b) provide and maintain the context (which includes a user-id) within which applications can run, (c) provide messaging authentication services, (d) spawn applications upon request of the user and (e) be able to create sessions for different users and switch between them. The Service & Content Discovery (SCD) component selects service and/or data components that satisfy the constraints of a specific request for functionality and/or content. It matches the specifications of the available components with the constraints

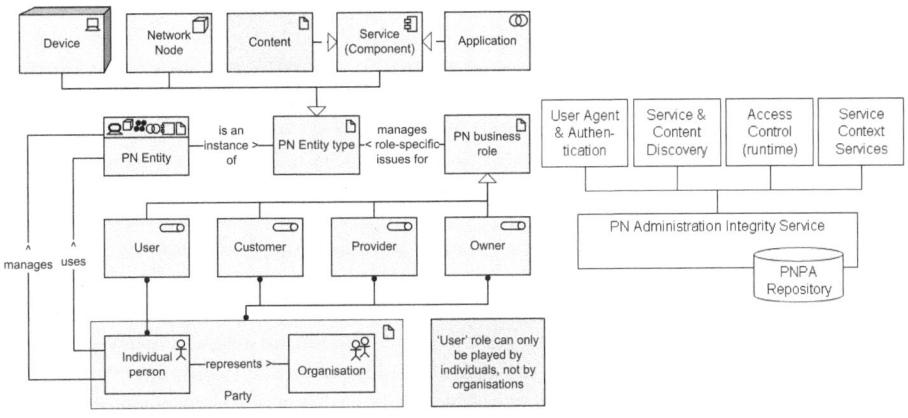

Fig. 1. PN architectural model (left), and functional blocks (right)

from the request, and passes the set of selected components to Access Control (AC). The AC component prevents the illegitimate use of services and/or content. Its task is to filter out service and/or content components from the set provided by the SCD, for use in the context of a specific application session. This is done by deciding for each component whether its execution/access is authorized, which means it must satisfy all constraints (rules) imposed by its Owner, Provider, Customer and User. The Service Context Service component enables services to enhance their efficiency and/or effectiveness by determining and using information from the context of the service relevant for the various PN Entities involved.

We implemented this architecture in a demonstrator as described in [5,6]. On the whole we conclude that the demonstrator works well. However, for commercial deployment a number of research issues still need to be tackled. For instance, what are the consequences of such architectures on the business processes of organizations such as IT departments of the work place? How should the PNPA and the rule engines be implemented? Centralized or distributed? How should the rule engines be constructed? How are security, privacy and ease of use in a PN managed in a balanced way? What algorithms and protocols are needed for supporting automated decision-making and follow-up? For example, PN content management requires intelligent, dynamic rules and protocols for replication, streaming, and deleting of content on the various PN Entities [7].

3 Conclusions and Future Work

The PN service layer architecture and the PN network layer architecture complement each other in the sense that the PN service layer assumes PN networking functionality, and the PN network layer is only useful if something is actually done with the messages it sends around. The service layer architecture specifies functionality for runtime selection and execution of appropriate service components available in the PN, resulting in a highly dynamic, personalized, and context-aware provisioning of PN services to the user.

The service layer also imposes new requirements, for example with respect to the development of service components. Developers of reusable service components for PNs may no longer rely on assumptions with respect to the applications interacting with their service component. As a result, high quality PN service components are much more sophisticated than the classical, silo-oriented application programs we often find in mobile phones.

We experienced that a pure device-oriented approach to PNs inevitably leads to a lot of discussion about the role of commercial operators and service providers in realizing and providing PN services to users. The PN service layer architecture provides a firm base to solve these issues. The architectural model clearly connects the responsibilities of the various business roles with the individual properties (resources) of the PN Entities involved. Any desired PN service component and/or its management that is too taxing for the device or too complex for the user to manage, may be provided by a centralized server instead.

It is obvious that achieving universal Personal Networking is highly dependent on standardization of mechanisms, protocols, data models, etc. of the various PN Entity Types involved. Unfortunately this may concern more or less every standards development body in the world, given the degree of heterogeneity that is inherent to PNs. For the standardization of the service layer components we are currently targeting the Open Mobile Alliance, which recently started a Converged Personal Networks Services work item. There we are advertising the service-oriented approach [8], which is well adopted by many partners.

References

1. Prasad, R.: Personal Networks and 4G. In: 49th International Symposium ELMAR-2007 focused on Mobile Multimedia, pp. 1–6. IEEE Press, New York (2007)
2. Niebert, N., et al.: Ambient Networks: an Architecture for Communication Networks beyond 3G. IEEE Wireless Communications 11(2), 14–22 (2004)
3. Niemegeers, I.G., Heemstra de Groot, S.M.: From Personal Area Networks to Personal Networks: A User Oriented Approach. Wireless Personal Communications 22(2), 175–186 (2002)
4. Service Requirements for Personal Network Management (PNM) Stage 1. Technical specification TS 22.259 V9.0.0, 3GPP (2008)
5. Hillen, B., Jager, E., Joosten, R.: Detailed PN Management Functionality. Technical report Freeband/PNP2008/DA.2.3, Freeband PNP2008 (2008)
6. Joosten, R.: Light-weight Security Services for an Integrated Personal Network Demonstrator. Technical report Freeband/PNP2008/D2.13v1.0, Freeband PNP2008 (2007)
7. den Hartog, F.T.H., et al.: First Experiences with Personal Networks as an Enabling Platform for Service Providers. In: 2nd International Workshop on Personalized Networks, pp. 1–8. IEEE Press, New York (2007)
8. Selgert, F., den Hartog, F.T.H.: On the Definition and Architectural Needs of Personal Networks. Contribution OMA-TP-CPNS-2008-0032-INP_CPNS, Open Mobile Alliance (2008)

Connecting the Islands – Enabling Global Connectivity through Local Cooperation

Janus Heide, Morten V. Pedersen, Frank H.P. Fitzek, and Torben Larsen

Aalborg University, Aalborg, Denmark, Dept. of Electronic Systems
{speje,mvpe,ff,tl}@es.aau.dk

Abstract. In this work we consider the interconnection of islands formed by neighbouring devices in a highly dynamic topology. To allow for high mobility we take offset in a purely wireless infrastructure where all devices incorporate a global and a local wireless communication interface. We consider cooperation as a means to improve the utilization of these interfaces. Furthermore we reflect on the use of network coding as a technique for decreasing the complexity of cooperative strategies and presents obstacles that must be addressed before network coding can be deployed.

Keywords: Cooperation, Network Coding, Ubiquitous Networks, Wireless Networks.

1 Introduction

Several future visions on computer network usage and accessibility include the concept that the usability of most things can be improved through the use of an ubiquitous network connection. In particular a connection that is always available and which require less interacting from the user. Such a vision imposes new requirements on the underlying network architecture and may require that networks are deployed differently than what is the norm today.

If devices are required to maintain an autonomous and continuously connection the only solution is to incorporate some form of global wireless communication interface. We note that the notion of a global connection is relative, for example a mobile phone can be considered as globally connected, but there will of course be locations where there is no coverage. This holds for all communication systems. Typically one drawback of a long range wireless communication system is the relative low capacity. In most current mobile phones this issue has been mitigated by incorporating a high speed short range wireless and/or wired communication interface. Thus the user can use the local connection when available and is at the same time always guaranteed service through the global connection although at a reduce speed. Thus the premises in this work is that all/most devices have one fast short range interface and one slow long range interface. Figure 1 shows three islands of devices with global and local communication links. The devices that form these islands will not remain stationary and thus the map of the islands will constantly change; new islands will appear,

C. Hesselman and C. Giannelli (Eds.): Mobilware 2009 Workshops, LNICST 12, pp. 135–141, 2009.

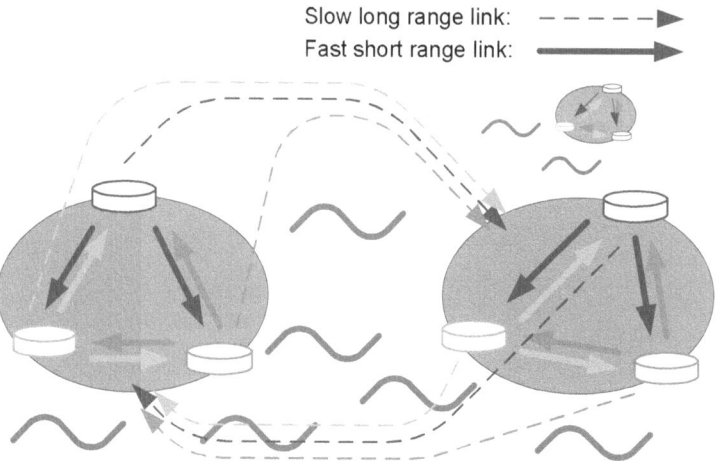

Fig. 1. Devices close to each communicates via their local interface and form an island, the global interface is used for communication with devices further away on other islands

islands will drift or change shape, and islands will disappear. This underlines the need for a wireless infrastructure as wired networks does not allow the necessary mobility.

In this type of wireless topology resources such as spectrum and energy are typically limited as the wireless medium is shared and devices are battery-driven. To use these resources more efficiently we can deploy cooperative strategies as they can take advantage of the differences between the local and global links. Cooperation can increase the throughput in cases where all devices on one island do not perform global communication simultaneously or wish to send or receive correlated information. With cooperation idle global links can be shared with neighbouring devices and thus the spectrum resource can be better utilized. If devices need correlated information they can share the burden of receiving the content and exchange their part of the information afterwards. Cooperation can also be used to reduce energy consumption. Consider the case where the standby power of the global link is significantly higher than for the local link. Here one device may take up a leader role and the burden of listening on the global link while the other devices turn off their global wireless link and thus conserve energy. If the leader detects that data is being transmitted to one of the devices with turned off global interface it may wake up that device via the local link.

Many such cooperative techniques have already been investigated and proven [1]. Thus the use of cooperation seems like a promising candidate for improving the communication. A cooperative system introduces a number of challenges and problems especially if the network topology is very dynamic.

2   Cooperation

The properties of the global channel influence how cooperation should be performed. In particular if the link of the different devices are independent or not.

If the global links are not independent devices on the island should not transmit simultaneously as the transmissions will otherwise collide. In this case cooperation can be used to distribute the burden of long range transmission among the devices on the sending island and thus share their energy resource. Alternatively the throughput can be increased if all devices on the receiving island overhear the transmissions from the source. If the sink losses a packet it is possible that one of the other devices on the island received it and in that case the sink may fetch the packet via its fast local connection. This form of cooperation is typically referred to as forwarding.

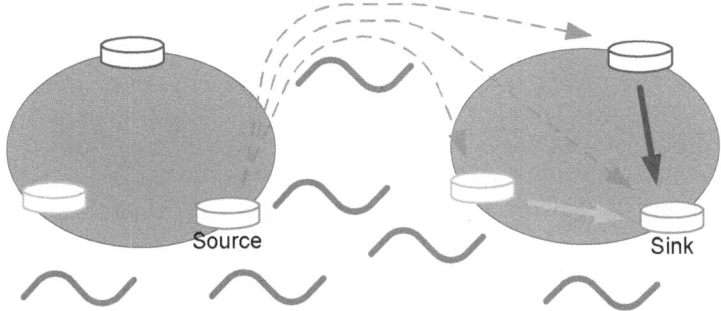

Fig. 2. The source transmits to the sink and the neighbouring devices of the sink cooperate with the sink

If the global links are independent the devices on the sending island may aid the source in transmitting in order to increase the throughput. One approach is to let the source spread the data to all devices on its island via its fast local interface. All devices on the island can then transmit parts of the data simultaneously and thus speedup the transmission. On the receiving island different devices can receive different parts of the data and forward it to the sink. This form of cooperation is often referred to as content-splitting.

Such cooperative approaches seems interesting as they potentially offer significant benefits e.g. increase in the transmission speed, increased reliability, and reduced energy consumption. However they also pose a number of challenges. One important problem is that all nodes need information about the topology, e.g. the source needs to know how to split the data and where to transmit it. In a simple network this type of knowledge can be maintained with a relatively low overhead but as the size and complexity of the topology increases so does the overhead. To remove the need for this information and reduce the complexity of the control system that guards the cooperation we consider the use of network coding.

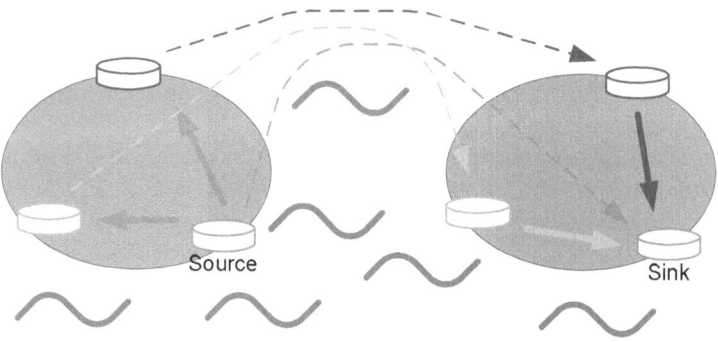

Fig. 3. The devices on the island of the source and the sink communicate cooperatively

3 Network Coding

Network coding is a network paradigm that allow for a completely different view of the flow of information in a network [2,3]. Network coding can be implemented in several different ways [4,5] and used for several different applications [6]. We consider linear network coding based on Galois field arithmetic as it is the most commonly used and well developed approach. Here we first provide a slightly simplified view on network coding to enhance readability and sum up important properties. The interested reader is referred to [7,8] for an introduction to network coding. Afterwards we explain how network coding may be beneficial for the; complexity, throughput, reliability, security and interoperability in the presented scenarios.

In a traditional network data to be send from a source via some network to a sink is divided into a number of packets N and transmitted one at a time. This approach is called store-and-forward or routing and as the name suggests nodes in the network receive packets and forward these without modifying the content of the packets. If reliable transmission is required each of these packets must be received and acknowledged by the intended sink. If not *all* N packets are received at the sink the source must retransmit the lost packet.

In network coding some additional operations are introduced namely; encoding, recoding, and decoding. At the source the original data is divided into the same N packets as in the traditional approach. However from these N packets new packets that are combinations of the N original packets can be created or *encoded*. The number of unique packets that can be encoded is generally much higher than N. The source can then transmit these packets to other nodes in the network which may in turn *recode* and forward any received *coded* packets. *Recoding* at a node in the network is similar to the initial coding at the source, however the resulting coded packets are only combinations of the subset of packets the node in question holds. In order for the sink to *decode* the original data it must receive *any* N independent coded packets and inverse the performed encoding operations.

Complexity: Network coding can significantly simplify the error correction system as the sink only need to receive N independent coded packets instead of N specific original packets. This approach can be used to reduce the complexity of both the intra- and inter-island communication.

Throughput: It has been shown that network coding satisfy the Max-flow min-cut theorem even in cases where traditional store-and-forward does not [2]. This theorem states that the maximum achievable flow in a network is limited by the bottleneck and thus there exists a network coding solution for all networks that achieves the maximal theoretical throughput. We cannot hope to do any better than this but note that such a solution may not always be easy to identify in a distributed manner.

Reliability: Network coding provides reliability by spreading the data maximally in the network. Therefore if a node fails the transmission can continue unaffected although at reduced speed.

Security: As transmitted packets are coded network coding naturally becomes an interesting concept when security in wireless networks is considered. If only the source and sink knows how the data was encoded no other nodes can decode the data. However the two end nodes are still faced with the traditional key exchange problem.

Interoperability: Network coding has been proposed for implementation on all layers of the network stack. However so far implementations have only been developed for the data link, transport, and application layer [4,9,10]. Implementing network coding on the application layer allows for deployment in existing networks agnostic to the underlying technologies. Thus network coding can be deployed without concern for interoperability as long as a common protocol specification is agreed upon.

Network coding provide many benefits however it also presents us with its own set of challenges of which we will consider some related to practical implementation.

4   Challenges and Considerations

Different scenarios and applications possess different requirements and thus the chosen network coding solution should reflect this. Data is coded in network coding and thus we must ensure that the coding throughput is sufficiently high which have proved to be a nontrivial problem [11]. In linear network coding primarily two properties guards the performance characteristics namely the *field size* and the *generation size* [7]. The field size specifies the size of the underlying Galois field and generally the complexity of the coding operations increases with the field size. The generation size specifies how many packets are combined into one coded packet and thus as more computations are needed per packet the throughput decreases as the generation size increases. In addition the generation size influences the delay at the sink because all data in one generation must be decoded before any of it is useful. Thus the minimum achievable delay is increased from one packet to the generation size. Therefore in applications such as

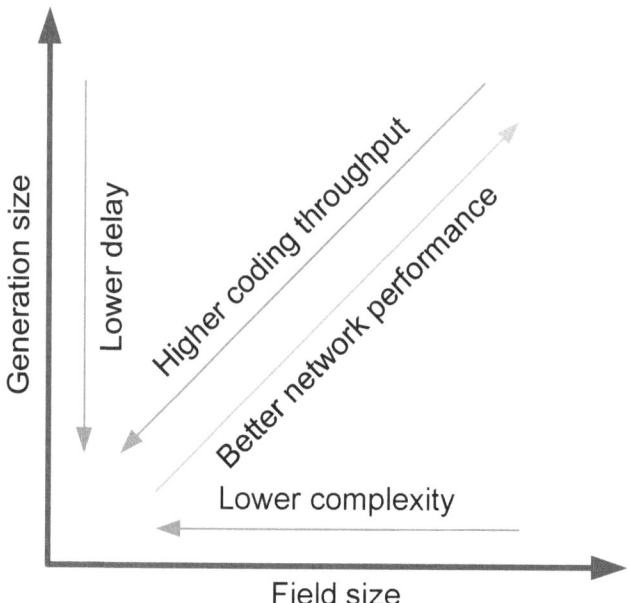

Fig. 4. Tradeoffs depending on the field and generation size

video or audio streaming the delay requirement limits the acceptable generation size whereas in file transfers this is not the case. This indicates that a low field size and generation size should be used, unfortunately this contradictions with the network performance which increases with the field and generation size [12], see Figure 4. We note that if we choose the lowest possible field and generation size linear network coding degenerates into broadcasting.

Additionally the coding throughput will influence the energy consumption because as performance of the coding decreases the CPU utilization increases which leads to increased energy usage.

This work has presented an approach for interconnection of ubiquitous island through the use of cooperation and network coding. We have briefly presented simple cooperation approaches and explained how communication among islands can be improved through cooperation. To decrease the complexity of cooperative strategies we consider network coding and point out some of the benefits it provide. Finally we reflect on some of the problems that must be solved before network coding can be deployed.

References

1. Fitzek, F., Katz, M. (eds.): Cooperation in Wireless Networks: Principles and Applications – Real Egoistic Behavior is to Cooperate!, April 2006. Springer, Heidelberg (2006)
2. Ahlswede, R., Cai, N., Li, S.Y.R., Yeung, R.W.: Network information flow. IEEE Transactions on Information Theory 46(4), 1204–1216 (2000)

3. Yeung, R.W., Zhang, Z.: Distributed source coding for satellite communications. IEEE Transactions on Information Theory 45(4), 1111–1120 (1999)
4. Katti, S., Rahul, H., Hu, W., Katabi, D., Medard, M., Crowcroft, J.: Xors in the air: practical wireless network coding. In: Proceedings of the 2006 conference on Applications, technologies, architectures, and protocols for computer communications (SIGCOMM 2006), September, 11-15, pp. 243–254. ACM Press, New York (2006)
5. Ho, T., Koetter, R., Medard, M., Karger, D., Ros, M.: The benefits of coding over routing in a randomized setting. In: Proceedings of the IEEE International Symposium on Information Theory, ISIT 2003, June 29 - July 4 (2003)
6. Fragouli, C., Soljanin, E.: Network Coding Applications. Now Publishers Inc. (January 2008)
7. Fragouli, C., Boudec, J., Widmer, J.: Network coding: an instant primer. SIGCOMM Comput. Commun. Rev. 36(1), 63–68 (2006)
8. Ho, T., Lun, D.S.: Network Coding An Introduction. Cambridge University Press, Cambridge (2008)
9. Katti, S., Katabi, D., Hu, W., Rahul, H., Medard, M.: The importance of being opportunistic: Practical network coding for wireless environments. In: Proceedings of 43rd Allerton Conference on Communication, Control, and Computing (2005)
10. Sundararajan, J.K., Shah, D., Médard, M., Mitzenmacher, M., Barros, J.: Network coding meets tcp. CoRR abs/0809.5022 (2008)
11. Heide, J., Pedersen, M.V., Fitzek, F.H., Larsen, T.: Cautious view on network coding - from theory to practice. Journal of Communications and Networks, JCN (2008)
12. Heide, J., Pedersen, M.V., Fitzek, F.H., Larsen, T.: Network coding for mobile devices - systematic binary random rateless codes. In: The IEEE International Conference on Communications (ICC), Dresden, Germany, June 14-18 (to appear, 2009)

Mapping the Physical World into the Virtual World: A Com2monSense Approach

Richard Gold, Vlasios Tsiatsis, Jan Höller, and Vincent Huang

Ericsson Research

Abstract. Networks of the future will contain a vast array of sensors and actuators which users will wish to interact with. Orchestrating this task effectively and efficiently will be one of the major challenges facing the deployment of these future networks. We propose a three-tier architecture which provides a middleware to resolve high-level service requests from applications to low-level sensor or actuator actions. Our architecture is designed around the core principle of horizontalization, that is, breaking apart vertically integrated silos into their component parts, thus allowing flexible recombination and reuse of data & services.

1 Introduction

Research in SANIs (Sensor Actuator Network Islands), i.e., a group of sensors with networking capabilities communicating with a gateway, has typically been concerned with issues such as: radio design, operating systems, network protocols and energy efficiency of highly specialized (e.g. enterprise, environmental) and small to medium scale sensor network islands [5]. The focus of this research is more on the deployment issues (e.g. coverage, energy efficiency [4] etc.) and less on the purpose of the deployment of sensor networks which is to provide an information interface from/to the physical world. This research has been extremely useful, but its narrow focus necessarily requires that the larger picture of how sensor networks are used has not been equally as well addressed. Specifically, the actual applications, use cases and scenarios in which sensor networks will be used will be a major influence in the design of sensor networks. Additionally, the networks of the future will be deeply heterogeneous (Section Assumptions) which will affect the design of any architecture for usage in such networks. In this paper we present an alternative outlook which focuses on how sensor networks will actually be used. Our research agenda is not focused on traditional areas of sensor networking research as outlined above. We are, however, primarily interested in an infrastructure that ties together the multitude of sensor and actuator deployments and exposes the collective set of services to be used by applications. The main contribution of this paper is the design, analysis and prototype of such an infrastructure constructed around the design principle of horizontalization by which we mean a horizontal market where the services offered by the sensor networks can be combined and reused by many different apapplications. This approach can also be thought of as a middleware for sensor networks. Currently sensor networks are typically deployed in vertically integrated silos which are self-contained. This means that if new functionality is required, then an entirely new silo must be deployed. Silos also represent information separation which

C. Hesselman and C. Giannelli (Eds.): Mobilware 2009 Workshops, LNICST 12, pp. 142–146, 2009.
© ICST Institute for Computer Sciences, Social-Informatics and Telecommunications Engineering 2009

is in contrast with the information aggregation and and fusion that is desirable from sensor actuator network islands. We see this as inefficient and one of the main goals of our architecture is to allow the reuse and re-combination of the data and services of sensor networks without requiring the deployment of new silos. We have already begun preliminary implementation work based on Middleware and Semantic Web technologies. We believe that our architecture is applicable to a wide range of scenarios and we describe and analyze a carefully selected example.

2   Com2monSense

The Com2monSense architecture is based on the middleware approach outlined above and was partially presented in [1,3,2]. The Middleware layer is the central point of the system, providing a natural point for the sensor networks and the applications to connect to. Additionally, it also provides an opportunity to bring 3rd party services into the interactions between the sensor networks and the applications, potentially providing services such as maps and tag resolvers for RFID. One of the direct consequences of this approach is the breaking up of the vertical silos of the Client/Server architecture into its separate components. This means that the data and services of the sensor networks are now re-combinable. Another consequence is enabling the creation of new functionality without deploying an entirely new silo.

One possible realization of the architecture described above is shown in Figure 1.

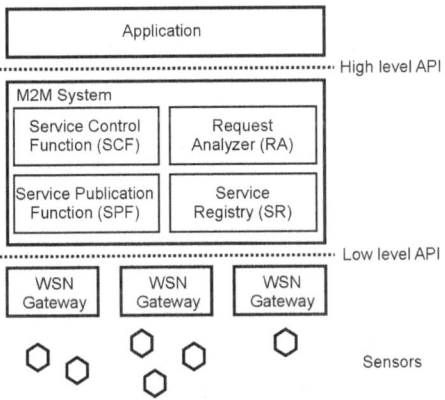

Fig. 1. High-level System Overview

The main components of the system are as follows:

Application: The application resides on the client device and is responsible for representing the clients' preferences and performing the user's desired actions. The application sends high-level service requests to the M2M system.

The M2M System: This is comprised of the following parts:

- **Service Control Function (SCF):** This is the main multiplexing & demultiplexing point of the M2M System. High-level service requests come in to the SCF from the application and low-level function call responses from the sensor network gateways come in to the SCF. The SCF then needs to multiplex the service requests to the appropriate part of the M2M System and demultiplex the function call responses to the correct part of the M2M System.

- **Request Analyzer (RA):** The RA is the key part of the M2M System. It is responsible for breaking down the high-level service requests into low-level function calls which are to be sent to the gateways.
- **Service Registry (SR):** The SR keeps track of which sensor networks are connected to the M2M System, which capabilities they have and how to reach them (e.g., which network technology is used to communicate with them).
- **Service Publication Function SPF:** This is a module for future use and we do not consider it further in this paper.

Gateways: The gateways manage the attachment and subsequent detachment of sensors and the general network configuration. The sensors will report their capabilities to the gateway which will then report the set of combined capabilities of its attached sensor networks to the M2M System as a set of services.

Sensors: The sensors register at the gateway automatically when they are powered on and provide a list of their capabilities. This should be performed with some standardized auto-configuration protocol. They then either periodically report information to the gateway or respond to explicit requests forwarded to them via the gateway according to how they are configured.

3   Implementation

In order to realize the Com2monSense architecture, a variety of technologies need to be employed. Since we are aim to support multiple applications and sensor networks, it is critical that we have a standardized set of interfaces:

- Interface from the application to the middleware (for example, SOAP, SIP or XMPP)
- Interface between 3rd party providers (for example, AJAX)
- Interface from the middleware to the sensor networks (For example, SIP or XMPP)

Once these are in place, the next step is the creation of the middleware itself. The primary goal of the middleware is the translation of high-level user queries to low level SANI calls by combining information from multiple sources to present a coherent view to the user. We believe this to be the part of the architecture that requires the most work. Currently we are looking at Semantic Web technologies to assist us in this area. A goal of the Semantic Web to make information machine-processable is also useful for helping us resolve high-level service requests to low-level sensor network function calls. Finally, the goal of the Semantic Web to integrate data from heterogeneous sources makes it a natural fit for our ambition to combine data from multiple sensor networks for consumption by multiple applications.

We use the ontology capabilities of the Semantic Web to represent the various aspects of the physical world. For example, sensors in different rooms, cars, streets, buildings, etc. can be represented in an ontology with explicit links between them. The motivation for having these links between the representation of sensors is that we wish to use *inference* as method of working out which sensors can be used to satisfy which service requests, i.e., working out how to break down high-level service requests to low-level function calls.

4 Scenario: Smart Open Spaces

We envisage the instrumentation of public, open spaces with various types of sensor to enable a rich set of services for the public. This will take the form of mapping the physical space into the Internet by allowing the public to interact with the physical environment through the Internet via the use of mobile devices. Interaction with the local environment can take place both locally and remotely. It will be possible for the virtual representation of the physical environment to be queried, annotated and remodelled by each individual user. The user will be able to discover the physical state of the environment (weather conditions, etc.) as well as access services such as gaming, tourism and community services. When a member of the public visits the smart open space, they may wish to know in advance the current status of the open space. For example, how many people are already there? What is the weather like? If the user then proceeds to visit the open space, they may wish to view annotations left by other people regarding Points-of- Interest and nearby services such as restaurants or museums.

The smart open space provides a horizontalized system where multiple service providers can attach to the open space and advertise their services and information models which users can then selectively choose or enable. This scenario represents an enhancement of open spaces for the general public. It also provides a neutral platform for companies to provide services and/or information. It will make it easier and quicker for the general public to discover better information about their physical environment and also use a rich array of services to interact with the environment. Due to the face that information is being collected about people's movements in public spaces, privacy is a major concern.

5 Conclusions

We have presented a three tier architecture for the processing and presentation of sensor data called Com2monSense. Our architecture is based around the core design principle of horizontalization, that is, breaking apart vertically integrated silos into their component parts thus allow recombination and reuse of the data and services of sensor networks. For future work we plan to continue our prototyping efforts to create a robust implementation of our architecture and examples applications conforming to the presented scenarios. Also, we plan to analyze in-depth the issues of privacy and security which we believe are essential to widespread uptake of such an approach.

References

1. Krco, S., Johansson, M., Tsiatsis, V.: A CommonSense Approach to Real-world Global Sensing. In: Proceedings of the SenseID: Convergence of RFID and Wireless Sensor Networks and their Applications workshop, ACM SenSys 2007, Sydney, Australia (November 2007)

2. Huang, V., Johansson, M.: Usage of semantic web technologies in a future M2M communication system. In: Proceedings of the 1st European Semantic Technology Conference, Vienna, Austria, May 31- June 1 (2007)
3. Huang, V., Javed, M.K.: Semantic Sensor Information Description and Processing. In: Proceedings of the Second International Conference on Sensor Technologies and Applications, SENSORCOMM 2008, Cap Esterel, France, August 25-31 (2008)
4. Hadim, S., Nader, M.: Middleware Challenges and Approaches for Wireless Sensor Networks. IEEE Distributed Systems Online (2006)
5. Römer, K., Friedemann, M.: The Design Space of Wireless Sensor Networks. IEEE Wireless Communications (December 2004)

Ubiquitous Mobile Awareness from Sensor Networks

Theo Kanter, Stefan Pettersson, Stefan Forsström,
Victor Kardeby, and Patrik Österberg

Mid Sweden University, 851 70 Sundsvall, Sweden
{theo.kanter,stefan.pettersson,stefan.forsstrom,victor.kardeby,
patrik.osterberg}@miun.se

Abstract. Users require applications and services to be available everywhere, enabling them to focus on what is important to them. Therefore, context information (e.g., spatial data, user preferences, available connectivity and devices, etc.) has to be accessible to applications that run in end systems close to users. In response to this, we present a novel architecture for ubiquitous sensing and sharing of context in mobile services and applications. The architecture offers distributed storage of context derived from sensor networks wirelessly attached to mobile phones and other devices. The architecture also handles frequent updates of sensor information and is interoperable with presence in 3G mobile systems, thus enabling ubiquitous sensing applications. We demonstrate these concepts and the principle operation in a sample ubiquitous Mobile Awareness service.

Keywords: context-aware applications, mobile systems, wireless sensor networks.

1 Introduction

People are expecting applications and services to provide value to them, such that they can focus on what is important in life. Recently, vendors of mobile devices (e.g., mobile phones, smartphones, PDAs, etc.) have started to include sensors in these devices (e.g., accelerometers, GPS, photo detectors, short range radio, etc.). Including sensors in a user device offers more information about what the user is doing or what situation the user is in to local applications, i.e., so-called context information. In this case users or applications lack means to share context information with remote users or applications.

Wireless sensors and sensor networks (WSNs) are important sources for context information, for which solutions exist to collect sensor data via gateways, for storage in servers, such as for monitoring the environment. These solutions are an insufficient response to the need for users to share context information in a distributed way.

Earlier work has focused on brokering of sensor information with web services via a 3G mobile system, such as Mobilife [1], brokering of sensor information via SIP servers [2], brokering of sensor information via clients in end-devices (opportunistically) connected with web services to servers on the Internet [3-5], and finally, other approaches offer gateways to mobile system for aggregating information from WSNs and making this available in services [6,7].

C. Hesselman and C. Giannelli (Eds.): Mobilware 2009 Workshops, LNICST 12, pp. 147–150, 2009.
© ICST Institute for Computer Sciences, Social-Informatics and Telecommunications Engineering 2009

Krco et al. presented an approach [7] providing gateways to mobile systems for collecting data from sensor networks. Sensor nodes lack identities assigned to a mobile device attached to a 3G mobile system or applications running in mobile devices register with a 3GPP IP-Multimedia Subsystem (IMS).

The above solutions are unsatisfactory answers to the challenge that we have identified, which is to provide means for the spontaneous sharing of context information from any locally available sensor with remote users or applications. The purpose of the required solution is to enable the spontaneous orchestration of local service behavior based on context information from locally and remotely available sensors.

In Europe, 3G offers ubiquitous coverage with packet data services (GPRS) via various radio access networks (GSM, EDGE, W-CDMA, HSPA). Thus mobile users can access both services on the Internet as well as services that run on the 3GPP IP Multimedia Subsystem (IMS). We argue that the solution should build on the ubiquity of wireless broadband connectivity for sharing of context between remote users, devices, and applications. In line with this, we present an architecture with support for ubiquitous provisioning and sharing of context information originating from sensors and WSNs via mobile systems. In our approach, end-devices share sensor information both via the Internet and via support in mobile systems.

2 Exchanging Sensor Information with 3GPP IMS

Aggregation of information from sensor networks in mobile devices and provisioning of context data in our support occurs via agents co-located with SIP end points, such as available in mobile phones. Mobile devices exchange information with presence support via SIMPLE [8] signaling using GPRS via 3G access.

We require real-time response times in provisioning of context data on the Internet, mandating the exchange of context data via a P2P infrastructure, which employs the Distributed Context eXchange Protocol (DCXP) [9] which

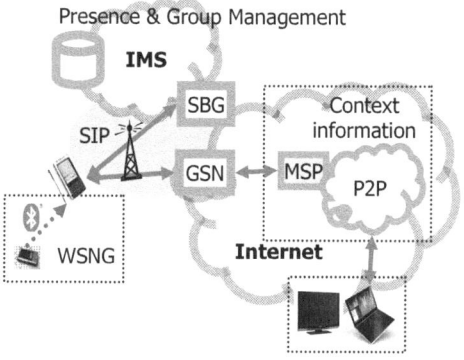

Fig. 1. Overview of the architecture

coexists and cooperates with 3G IMS (see Fig. 1). DCXP adheres to principles of SIMPLE for event notifications and utilizes an underlying Distributed Hash Table (DHT) as a registry for references to context information. Context sharing via DCXP circumvents severe limitations with IMS presence [10] in that the latter (a) causes long response times, (b) imposes a data model which causes excessive overhead when searching and browsing context information, and (c) has scalability problems due to centralized service access.

Sensor agents running in mobile phones connect to and register with a Mobile Sensor Proxy (MSP) in order to participate in the P2P network. Sensor agents also connect to IMS presence in order to make context information available as presence data. The MSP shields the P2P network from the behavior of the radio link, by shielding it from packet loss and other possible disruptions of the communication with the mobile. Nodes in the P2P network with ample resources may offer persistent storage of context information and to perform aggregation and filtering tasks in search and browse operations.

3 Ubiquitous Mobile Awareness Services

We built a prototype of our system and developed a Ubiquitous Mobile Awareness (UMA) application on top of it. The UMA application currently visualizes sensor information related to objects in Google Maps and also alerts the user to changes such as passing of other objects belonging to the user's group. The icon of each mobile user can expand to display sensor values, see Fig. 2.

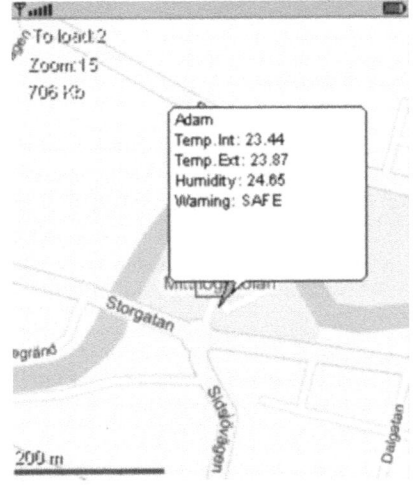

Fig. 2. The user interface of the UMA application

Using this service, you may position members in groups and query their presence profiles regarding associated sensor data from IMS. With a PC, you may also utilize the extra computing power and larger display to create views with more objects and sensors.

For the design, we used the Ericsson Mobile Java Communication Framework [11], which offers java libraries with SIP support according to JSR281, to create MIDlets that run on top of Sony Ericsson C702, or other Java ME capable mobile phones. We built a wearable bridge to associate WSNs with a mobile phone or any other mobile device with a Bluetooth interface. Our current prototype employs mobile phones which read values from humidity and temperature sensors and forwards these to both a presence server in IMS and a Mobile Sensor Proxy. The mobile phones may also utilize the built-in GPS for positioning.

The wearable bridge can easily be extended to hold more than humidity and temperature sensors, e.g., compass, accelerometers, CO_2, and radiation. Availability of such information enables monitoring of environmental parameters and traffic conditions, etc., and emphasizes the important of sharing of sensor information for creating ubiquitous awareness in on-line communities and mobile applications.

4 Conclusions

In this paper we presented an architecture for the sharing of context information derived from sensor networks wirelessly attached to mobile phones and other devices. The architecture handles frequent updates of sensor information in real time and is interoperable with presence in 3G mobile systems, thus enabling ubiquitous sensing applications in mobile devices for groups and individuals. Our future work will focus on scalable real-time provisioning and distributed storage of multi-dimensional sensor information and exploring its utility in ubiquitous mobile awareness scenarios.

Acknowledgments

The research is partially supported by the regional EU target 2 funds, regional public sector, and industry such as Ericsson Research and Telia Sonera.

References

1. Klemettinen, M. (ed.): Enabling Technologies for Mobile Services: The MobiLife Book. John Wiley and Sons Ltd, Chichester (2007)
2. Angeles, C.: Distribution of Context Information using the Session Initiation Protocol (SIP). Master of Science Thesis, KTH ICT school, Stockholm, Sweden (2008)
3. Abdelzaher, T., Anokwa, Y., Boda, P., Burke, J., Estrin, D., Guibas, L., Kansal, A., Madden, S., Reich, J.: Mobiscopes for human spaces. Pervasive Computing 6(2), 20–29 (2007)
4. Hull, B., Bychkovsky, V., Zhang, Y., Chen, K., Goraczko, M., Miu, A., Shih, E., Balakrishnan, H., Madden, S.: Cartel: A Distributed Mobile Sensor Computing System. In: 4th international conference on Embedded networked sensor systems, pp. 125–138. ACM Press, New York (2006)
5. Grosky, W.I., Kansal, A., Nath, S., Liu, J., Zhao, F.: SenseWeb: An Infrastructure for Shared Sensing. Multimedia 14(4), 8–13 (2007)
6. Hjelm, J., Oda, T., Fasbender, A., Murakami, S., Damola, A.: Bringing IMS Services to the DLNA Connected Home. In: Pervasive 2008, 6th International Conference on Pervasive Computing (2008)
7. Krco, S., Cleary, D., Parker, D.: P2P Mobile Sensor Networks. In: HICSS 2005. Proceedings of the 38th Annual Hawaii International Conference on System Sciences, p. 324c (2005)
8. Rosenberg, J.: SIMPLE made Simple: An Overview of the IETF Specifications for Instant Messaging and Presence using the Session Initiation Protocol (SIP). Internet-Draft, IETF SIMPLE Charter (2008)
9. Chazalet, B.: Opportunistic Media Delivery: utilizing Optimized Context Dissemination via the Service Interface in Ambient Networks. M.Sc. Thesis, Royal Institute of Technology (2007)
10. Lin, L., Liotta, A.: A Critical Evaluation of the IMS Presence Service. In: 4th International Conference on Advances in Mobile Computing & Multimedia, pp. 19–28 (2006)
11. Mobile Java Communication Framework. Ericsson Labs Developer, released (November 2008), http://dev.labs.ericsson.net/ tiki-view_blog.php?blogId=2&first=1

Author Index